Hypothesis Testing Made Simple

By

Leonard Gaston Ph.D.

"The author is overwhelmingly grateful to Brenda Van Niekerk (brendavniekerk@hotmail.com) for the incredibly efficient and helpful manner in which she applied her impressive computer skills in putting this material into e-book format. Brenda, is was a distinct pleasure to work with you!"

"Appreciation is given to Dr. Ben Williams, a model of excellence as department chair at Central State University and a great mentor."

Cover Design by Laura Shinn Designs
http://laurashinn.yolasite.com

"Appreciation is extended to Laura Shinn (laurashinn.author@gmail.com) for her excellent cover graphics."

Copyright © 2014 Leonard Gaston Ph.D.
All rights reserved.

ISBN-13: 978-1500849993

ISBN-10: 1500849995

No part of this publication may be reproduced or distributed in any form or by any means (electronic, mechanical, photocopy, recording, or otherwise) or stored in a database or retrieval system, without the prior written permission of the publisher or the author.

Table of Contents

Introduction 7
Chapter 1 9
Chapter 2 20
Chapter 3 29
Chapter 4 35
Chapter 5 40
Chapter 6 56
Chapter 7 73
Chapter 8 80
Chapter 9 108
Chapter 10 127
Chapter 11 139
Chapter 12 153
Chapter 13 159

Introduction

Have you ever found the study of statistics difficult and the subject of hypothesis testing intimidating? Does your thesis or research project advisor want you to use a hypothesis test? This little book can help! In understandable, down-to-earth language it describes eight simple hypothesis tests. Among them we hope you can find one that you can use to make sense of the numbers collected in your thesis or other research project.

Study of this little book should help you proceed with confidence to gather usable data, select a suitable hypothesis test, and apply that test. Once you master its contents, you will know what you did and why you chose that particular test. Away with throwing a hodgepodge of numbers into a PC program and then wondering if you can explain the results! Be in a position to confidently pick your test, put it to use, and then explain your results.

This is not a full blown statistics book. It omits many areas covered in a typical tome. And it does not cover all possible tests – some are quite sophisticated. What it does attempt to do is give the non-mathematician a grounding in the basics of a few potentially useful tests – enough, we hope, to get most of us through our research projects.

*****

Note: Many hours have been spent in an attempt to correct typos and any possible errors in problem solutions, however the writer does not recall a text book that did not contain errors. Murphy's seventh law states that "Nature always sides with the hidden flaw". If you find it did, the writer asks you forgive any inconvenience this might cause.

Chapter 1

Numbers, Central Tendency, and Dispersion

What this chapter will do for you.

From this chapter you will learn about the four classes of data. More importantly, you will learn about measures of central tendency in a group of numbers, measures of dispersion, and how to calculate the mean and standard deviation of ungrouped and grouped data.

Four Kinds of Numbers

There are four classes of data, or kinds of numbers, each providing greater amounts of information and offering increased usefulness for analysis.

In what we could call the bottom class, we have **nominal data**. We can count things and put them in categories. A history class for example might contain ten men and twelve women. Or a pasture might contain four red cows, two black cows, and one yellow cow.

The next step up would be **ordinal data**. Here, things are ranked. One authority's ranking of football teams might rank Auburn over Alabama for instance. (Or it could be the other way around.) A ranking of the top ten teams would presume to rank them in order of excellence.

In the rankings above, assuming Auburn might rank above Alabama, how much better is Auburn? How many points for example? The rankings won't tell us. Such a measurement is provided by **interval data**. Although such information relative to football teams might well be suspect, there are instances where interval data is real and useful – a thermometer for example. On a thermometer the intervals between numbers are meaningful. It would take an input of just as many calories to raise a beaker of water from forty to fifty degrees Fahrenheit as is would to raise the temperature from fifty degrees to sixty degrees – or from sixty degrees to seventy degrees, and so on. We recognize of course that if the thermometer passes 32 degrees Fahrenheit on the way down, or 212 degrees on the way up (or whatever the boiling point of water would be at our altitude) a change of state would occur and at those points the calorie input would not be consistent.

The numbers we take for granted in most of our daily activities fall into the category called **ratio data**. Numbers that tell us how many pounds of dog food are in a bag, how many miles per gallon our cars get, or how far a track athlete can broad jump have one unique characteristic. Each has a meaningful zero point. Using ratio data we can add, subtract, multiply, divide, find square roots, and do hypothesis tests. Note: we will cover two tests that can be carried out with nominal data. Chapter 10 will introduce Chi (the "i" rhymes with the "I" in kite, not the "e" in scream) Square. Chapter 11 will describe the use of proportions. Chapter 13 will briefly discuss Rank Order correlation which can be used with ordinal data.

Measures of Central Tendency

Assume we have weighed nine ducks and obtained the following values in pounds (Note: we are working here with ratio data): 4,5,5,6,6,6,7,7, and 8. What would be the **arithmetic mean**, that is what we commonly call the "average"? We would find it by adding up the numbers (total = 54) and dividing by the number of ducks (nine). The average weight of the ducks is six pounds. Simply by looking at these numbers, as they are listed above, would likely lead us to guess this, without doing the calculation.

The value that appears most often in a group of numbers is called the **mode**. In the numbers above the mode would be six. The term is used outside the study of statistics where the meaning is similar: A fashion expert might say that the mode in ladies' coats this spring is blue – meaning that blue is the color seen most often.

The middle number in our group of duck weights, with just as many values below it as above it, would be six. This is called the **median**. In the numbers above the median is clearly seen to be 6. Suppose we had only eight ducks and the weights were 4,5,5,6,6,7,7, and 8. To get a number such that just as many weights were below it as above it, we would split the difference and say the median was halfway between the fourth number from the bottom (a 6) and the fourth number from the top (the other 6) and would still be six. On the other hand, if the eight numbers we had were 4, 5, 5, 6,7 ,7, 7, and 8 we would assume the median to be halfway between the 6 and the first seven, or six and a half.

In some situation, to illustrate the "average", the median of a group of numbers would be more useful than the arithmetic mean. Suppose there were sixteen houses in a township and a marketing service wished to publish a figure that more or less described the "average" value of houses in the township. Let's assume there are five appraised at one hundred thousand dollars, five appraised at a hundred fifty thousand dollars, and five appraised at two hundred thousand dollars. Up on a hill however, just barely within the township boundaries is one appraised at two million dollars.

If we compute the arithmetic mean we would find it to be $265,625. That would be misleading because all the houses except one would have been appraised for less than that figure – some for much less. In this case, the median price (half the houses above and half the houses below) would be $150,00 – a less misleading value.

Formulas for the mean, and their use.

We will now discuss in more detail the most used, and possibly most important, "average", the arithmetic mean, along with the weighted mean and the estimated mean of a frequency distribution. There is also a geometric mean which, in the interest of brevity, we will not address.

Formulas for calculating the means of ungrouped data (data that has not been grouped into classes) are given below, using the following symbols:

X is the symbol for a variable (a duck weight in this case).

$\overline{X}$ or "X bar" is the symbol for the arithmetic mean of a sample.

n stands for the number of values (variables, measurements) in a sample.

µ the Greek letter pronounced "Mu" is the symbol for the arithmetic mean of a population.

N stands for the number of values (variables, measurements) in a population.

Σ is the capital Greek letter Sigma, which stands for the summation or sum of a number of variables.

$\overline{X}_w$ is the symbol for the weighted mean.

W in the symbol above stands for "weight", that is, the number of times a particular value or variable appears in a group of numbers.

Let's assume that the first set of duck weights given above pertains to a sample of ducks taken from a larger flock and calculate the arithmetic mean of this sample.

$$\overline{X} = \frac{\Sigma x}{n}$$

or the sum of the weights of the ducks (54 pounds) divided by the number of ducks (nine), giving an average weight of six pounds. This is the calculation we did earlier when we intuitively added up the duck weights and divided by the number of ducks.

If our nine ducks were the entire population, the calculation would be the same but the symbols would be different.

$$\mu = \frac{\Sigma x}{N}$$

or the sum of the weights of the ducks (54 pounds) divided by the number of ducks (nine), giving an average weight of six pounds.

The number is the same but the symbols for the mean and for the number of variables are different because the first illustration is for a sample and the second for a population.

Using the formula below to calculate a **weighted mean** can be useful if there are many variables with the same value. In our examples here, with a small number of variables, calculating the weighted mean would be more trouble than it would be worth, but we can see that if we had a large number of variables, with some numbers repeated many times, the formula for the weighted mean might be easier to use.

The formula for the weighted mean is shown below. We will use it to calculate the weighted mean of our sample of ducks. (Notice: We will not use the Mu symbol. We will assume that when we calculate a weighted mean we are always using sample data.) The weight (w) of each variable is simply how many times it occurs.

$$\overline{X}_w = \frac{X_1W_1 + X_2W_2 + X_3W_3 + X_4W_4 + X_5W_5}{\Sigma W}$$

$$\overline{X}_w = \frac{(4)(1) + (5)(2) + (6)(3) + (7)(2) + (8)(1)}{1 + 2 + 3 + 2 + 1}$$

$$\overline{X}_w = \frac{(4) + (10) + (18) + (14) + (8)}{9}$$

$$\overline{X}_w = \frac{54}{9}$$

$$\overline{X}_w = \underline{6}$$

Notice that we have underlined our answer. This can be useful for identifying your problem solution once you calculate it.

A Frequency Distribution and the Mean of a Frequency Distribution

Why are data (numbers, variables) sometimes grouped into a frequency distribution? In some situations, large numbers of variables must be analyzed. As suggested above, if there are a number of identical values, using the formula for the weighted mean might be the simplest procedure. However, if the numbers are almost all different, perhaps because they have decimal values, use of a frequency distribution would be more efficient.

When a frequency distribution is constructed, upper and lower class limits are constructed for a number of classes, and the variables are sorted by how many would fit in each class. This is called the **frequency** of each class. Calculations can be simplified if the limits are chosen so that the midpoint of each class is an easy to handle number.

In the following example we will assume that we want to find the **estimated mean** of the following numbers. Remember our original nine duck weights above? Let's assume they were much more accurate than even pounds and were instead as follows:

4.05977, 4.57778, 5.49043, 5.51111, 6.00001, 6.30989, 6.99999, 7.12345, and 7.99999 pounds. We notice that no two are identical, so we can't use a weighted mean.

First we will simply treat these numbers as ungrouped data and find the mean.

$$\overline{X} = \frac{\Sigma x}{n}$$

$$\overline{X} = \frac{54.07242}{9}$$

$$X = \underline{6.00805}$$

This is the actual mean of these numbers. Now let us construct a frequency distribution (below) grouping these numbers into the classes of that frequency distribution. We will place each variable in its proper class, count how many variables are in each class, determine the representative variable for each class (the class midpoint), and then calculate the **estimated mean**.

If we look at our first variable (our first x) 4.05977, we see it fits between the class limits of 3.5 up to 4.5. It is the only number that fits there, so the frequency of that class is One. The midpoint of the class – halfway between the class limits is Four. So the frequency of that first class, reading from the bottom up, is 1 and the X or midpoint is 4.

The variables 4.57778 and 5.49043 fit into the next class. Its frequency (it has two variables that fit within that class) is 2. Its midpoint is 5.

5.51111, 6.00001, and 6.30989 fit in the third class. Its frequency is 3. We can see that the midpoint of that class is 6.

Similarly, 6.99999 and 7.12345 fit in the next class, and 7.99999 fits in the remaining class.

We have constructed a **frequency distribution**.

Class Limits	Frequency (f)	Midpoint (X)	(f)(X)	X²	(f)(X)(X) or (f)(X)²
7.5 up to 8.5	1	8	8	64	64
6.5 up to 7.5	2	7	14	49	98
5.5 up to 6.5	3	6	18	36	106
4.5 up to 5.5	2	5	10	25	50
3.5 up to 4.5	1	4	4	16	32
Column sums:	9		54	190	350

We can see that the sum of the frequencies is nine. This is the number of variables. The sum of the (f)(X) column is 54. This is the same number we calculated earlier when we summed up the variables individually. The number of variables in each class times the representative variable for that class (the midpoint of the class) – (f)(X) - is analogous to the (X)(W) terms for each variable when we calculated the weighted mean. The sum of this (f)(X) column is analogous to the numerator in the equation for calculating the weighted mean.

Note: The sum of the X^2 column is 190 and the sum of the f times x^2 column is 350. We will use those numbers later.

The formula for the estimated mean of a frequency distribution is as follows:

$$\overline{X} = \frac{\Sigma fx}{\Sigma f \text{ (or n)}}$$

$$\overline{X} = \frac{(1)(4) + (2)(5) + (3)(6) + (2)(7) + (1)(8)}{9}$$

It looks like our calculation of the weighted mean shown earlier doesn't it? And it is basically the same thing, only the representative variable for each class (the mid points of each class) is used. These midpoints correspond to the individual even variables of the weighted mean problem, and the frequencies shown for each class correspond to the weights used in the weighted mean problem. (You may notice that the frequency is written first, because that is the way the formula is stated, while in the weighted mean problem the variable is stated first, and then its weight, because the is the way the formula is commonly written.)

$$\overline{X} = \frac{\Sigma fx}{\Sigma f}$$

$$\overline{X} = \frac{54}{9}$$ The estimated mean of this frequency distribution is 6.

Dispersion is a term that means basically how spread out a group of numbers is. There are a number of ways to do this. We will define a limited number, and then home in on the **standard deviation**, which will be the most useful for hypothesis testing.

The range is simply the distance between the smallest value and the largest value. For example, in our original duck weights the smallest value was four and the largest was eight. The range was 8 – 4 or 4.

Sometimes a group of numbers is divided into Quartiles (four equal parts) and the limits of each are shown graphically.

The **mean deviation** of a group of numbers is sometimes used. Because the average deviation of the variables from the mean is shown, this lessens the impact of one or two or a few extremely large or small values located far away from the mean.

The measure of dispersion that we will focus on however is the standard deviation. If we assume a normal distribution of variables where the mean, median, and mode are the same, and the distribution is symmetrical around the mean, the distribution will look something like this.

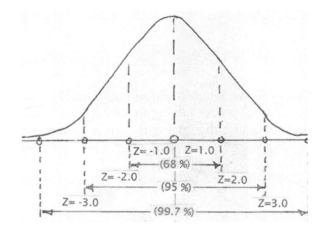

As we will discuss in more detail later the letter "Z" is a convenient abbreviation for the standard deviation. For example, we could measure to the right or left from the middle of this normal distribution of variables in units called Z's. If we measure out one Z we have measured out horizontally one standard deviation. The height of the curve shows how many variables of each possible size there are of that particular size. Please notice that the outer edges of the line tracing the distribution do not touch the base line in this illustration. Theoretically, they never do. We can visualize the line tracing the distribution as "floating" above the base line. It would get so close we could not detect a gap between them, but theoretically a gap would always be there.

We can see that, in <u>approximate</u> terms, a standard deviation could be thought of as about one sixth of the distance across a normal distribution. A widely dispersed distribution would have a large standard distribution. A compressed distribution would have a small standard deviation. Within the limits of three standard deviations to the right of the mean and three standard deviations to the left of the mean will reside 99.7 percent of all the variables in a normal distribution.

How is the standard deviation calculated?

Textbooks typically show the formula for calculating the standard deviation this way. You will notice that the mean of the distribution must be found, then the deviation (distance) of each variable from the mean must be determined. And then each of these deviations must be squared The sum of these squares is then divided by the number of variables. The formula shown is for a population.

$$\text{Standard deviation} = \sqrt{\frac{\Sigma(x - \mu)^2}{N}}$$

Note: For a sample standard deviation, which will be most useful in this book, the term N (the number in the population) is replaced by "n – 1", the number in the sample minus one.

A method many find simpler to use applies a formula which can be derived from the formula above. (If you are interested, that derivation is shown in the appendix to this chapter.) For the standard deviation of a sample, the corresponding formula would be as follows:

$$\text{Standard deviation} = \sqrt{\frac{\Sigma(x^2) - (\Sigma x)^2 / n}{n-1}}$$

To get values to plug into that formula we would need to build a table. Although we will have to square each variable to build the table, the total process using this formula to find the standard deviation will usually be found to be less time consuming than using the formula shown first. For simplicity we will use the original duck weights.

x	x²
8	64
7	49
7	49
6	36
6	36
6	36
5	25
5	25
4	16
54	336

$\Sigma x = 54 \qquad \Sigma x^2 = 336$

$$\text{Standard deviation} = \sqrt{\frac{(336) - (54)^2 / 9}{9-1}}$$

$$\text{Standard deviation} = \sqrt{\frac{(336) - (2916)/9}{8}}$$

$$\text{Standard deviation} = \sqrt{\frac{(336) - (324)}{8}}$$

$$\text{Standard deviation} = \sqrt{\frac{12}{8}}$$

$$\text{Standard deviation} = \sqrt{1.5}$$

$$\text{Standard deviation} = 1.2247$$

(Note: The enclosure drawn around the right side of the equation, called a "radical", shows that we will take the square root.)

Calculating the Estimated Standard Deviation from a table where data has been grouped into a frequency distribution.

Knowing how to calculate the standard deviation of a sample of variables will be necessary for doing the hypothesis tests described later. While we are calculating standard deviations, it would be good to see how we calculate the <u>estimated</u> standard deviation of the data in a frequency distribution. The end result will look a lot like the procedure we used to calculate the standard deviation of ungrouped data. First we must construct a **frequency distribution**.

Class Limits	Frequency (f)	Midpoint (X)	(f)(X)	X²	(f)(X)(X) or (f)(X)²
7.5 up to 8.5	1	8	8	64	64
6.5 up to 7.5	2	7	14	49	98
5.5 up to 6.5	3	6	18	36	108
4.5 up to 5.5	2	5	10	25	50
3.5 up to 4.5	1	4	4	16	16
Column sums:	9		54	190	336

Now we will need the last two columns of the table. The sum of the X² column is 190 and the sum of the f times x² column is 350. The formula we will use is similar to that used above.

$$\text{Standard deviation} = \sqrt{\frac{\Sigma(fx^2) - (\Sigma fx)^2 / n}{n-1}}$$

$$\text{Standard deviation} = \sqrt{\frac{336 - (54)^2 / 9}{9-1}}$$

$$\text{Standard deviation} = \sqrt{\frac{336 - (2{,}916)/9}{8}}$$

$$\text{Standard deviation} = \sqrt{\frac{336 - 324}{8}}$$

$$\text{Standard deviation} = \sqrt{\frac{12}{8}}$$

$$\text{Standard deviation} = \sqrt{1.5}$$

Standard deviation = <u>1.2247</u>

What Have You Learned and What Comes Next?

You now should have an understanding of the approximate meaning of the term standard deviation and have learned how the standard deviation can be calculated. If any doubt exists as to the calculations it might be worth the effort to go back and work through the examples given in calculation of means and standard deviations. The next chapter will introduce probability. Some familiarization with the concept of probability will be helpful in understanding hypothesis testing.

*****

Appendix to Chapter One. Derivation of the simpler-to-use formula for the standard deviation of ungrouped data.

$$\sigma = \sqrt{\frac{\sum (x - \bar{x})^2}{n - 1}}$$

$$\sigma = \sqrt{\frac{\sum (x - \bar{x})(x - \bar{x})}{n - 1}}$$

$$\sigma = \sqrt{\frac{\sum (x^2 - 2(\bar{x})(x) + \bar{x}^2)}{n - 1}}$$

$$\sigma = \sqrt{\frac{\sum (x^2) - \sum 2(\bar{x})(x) + \sum \bar{x}^2}{n - 1}}$$

$$\sigma = \sqrt{\frac{\sum (x^2) - 2(\bar{x})(\sum x) + n(\bar{x}^2)}{n - 1}}$$

$$\sigma = \sqrt{\frac{\sum (x^2) - 2(\sum x/n)(\sum x) + n(\sum x/n)^2}{n - 1}}$$

$$\sigma = \sqrt{\frac{\Sigma(x^2) - 2(\Sigma x^2/n) + (\Sigma x)^2 / n}{n - 1}}$$

$$\sigma = \sqrt{\frac{\Sigma(x^2) - (\Sigma x)^2 / n}{n - 1}}$$

Chapter 2

Probability

What this chapter will do for you.

This chapter will help you understand probability -- the likelihood that some event will occur under certain circumstances. Hypothesis testing is based on probability. An introduction to these basic fundamentals of probability can be helpful preparation for the study of hypothesis testing.

Probability refers to the likelihood that some event will happen. You might say that there is a fifty-fifty probability that you will be late for lunch because of work you must complete. You have established a subjective probability of .5 or fifty percent. If we flip a coin we know the probability of getting a tail is .5 or 50 %. A half of a pair of dice, a die, has six sides. If you roll that die you would expect a one in six chance of it coming to rest with any particular side facing up. (1/6 = 1.6667)

Four Probability Rules

Four rules are commonly used to calculate probabilities.

Probability of some event =

Two are called rules of addition because the first sign after the equal sign is an addition sign.

Probability of some event = ... +

Two are called rules of multiplication because the first sign after the equal sign is a multiplication sign

Probability of some event = ... * Note we are using a star or asterisk to indicate multiplication to avoid confusion between "x" which is often used as a multiplication sign and "x" signifying a variable.

We will learn to apply two rules of addition and two rules of multiplication.

What is commonly called the **General Rule of Addition** is used when two events are not mutually exclusive. (That means it is possible that both events could happen.) The **Special Rule of Addition** applies when two events are mutually exclusive.

The **General Rule of Multiplication** will be applied when two events are independent of each other. (That mean that the probability of one event is not affected by whether or not the other event occurs.)

The **Special Rule of Multiplication** will be applied when the probability of a possible second event is affected by whether or not some other event occurs.

In this area of probability, formulas are read somewhat differently than would be the case elsewhere in the use of algebraic formulas.

P(A or B) is read "The probability that event A will happen or event B will happen" or simply "The probability of A or B".

P(A and B) is read "The probability that event A will happen and event B will happen" or simply "The probability of A and B".

P(A) is read "The probability that event A will happen" or simply "The probability of A".

P(B) is read "The probability that event B will happen" or simply "The probability of B".

P(B/A) is read as "The probability that event B will happen given that event A has already happened" or simply "The probability of B, given A".

First Let us illustrate use the General Rule of Addition. That rule is stated like this.

P(A or B) = P(A) + P(B) – P(A and B)

The last term -- P(A and B) -- is called the joint probability. It seems logical that we would add P(A) and P(B), but why should the formula tell us to subtract the joint probability? This is to avoid double counting. Perhaps this illustration will help. You have a deck of cards. Let's say event A would be cutting the cards and turning a Heart face up. Event B would be reshuffling the cards, making a cut and turning up an Ace. Drawing on our knowledge of probability we would note that in a 52 card deck (no jokers) there are thirteen Hearts. So P(A) would be 13/52. We note that there are four aces in the deck so P(B) would be 4/52. But notice, we are counting the Ace of Hearts twice. We need to subtract this joint probability of 1/52.

Let's work the problem.

P(A or B) = P(A) + P(B) – P(A and B)

P(A or B) = 13/52 + 4/52 – 1/52

P(A or B) = 17/53 – 1/52

P(A or B) = 16/52

P(A or B) = .3077

If we were somehow uncomfortable with fractions we could have immediately converted each probability to a decimal. Due to rounding the final answer might not be exact, but it would be close.

P(A or B) = 13/52 + 4/52 – 1/52

P(A or B) = .25 + .0769 − .0192

P(A or B) = .3269 − .0192

P(A or B) = .3077 In this case early conversion to decimals and consequent rounding did not affect the answer.

Now to apply the Special Rule of Addition, where events are mutually exclusive. That rule is stated like this.

P(A or B) = P(A) + P(B)

Suppose we somehow know that the probability that you will sign up for history at eight o'clock Tuesday and Thursday is .4 and the probability that you will sign up for English at eight o'clock Tuesday and Thursday is .3.

P(A or B) = P(A) + P(B)

P(A or B) = .4 + .3

P(A or B) = .7

Next we will illustrate the use of the General Rule of Multiplication. Recall: It is used when two events are independent.

P(A and B) = P(A) * P(B)

Suppose a college has set commencement practice for today. If it rains today the practice is to be held tomorrow. If the probability that it will rain today is .4 and the probability of rain tomorrow is .7 what is the probability that it will rain both days and practice will be eliminated or will be held in the rain?

P(A and B) = P(A) * P(B)

P(A and B) = .4 * .7

P(A and B) = .28

Let's think about The General Rule of Multiplication for a moment in practical terms. Suppose you were parked in the boss's parking place, he was driving into the parking lot, and you ran out to move your car to the adjacent space which is now vacant. How many items on your car would have to function properly for you to do this. Let's say the each item has a 90 % or .9 probability of functioning properly. The ignition switch must work both to turn on the ignition and to activate the starting mechanism, a solenoid must function to engage the starter, the starter must turn, the ignition system considered as a whole must function, the fuel system must function, the transmission selector must function, the transmission must function, the steering mechanism must function sufficiently to back out, align the car, and pull forward, the brakes must function, and the selector must function to put the car in park, that function of the transmission must work, the brakes must work, and the ignition switch must function again to shut the engine down. We have just counted up thirteen things that need to work to complete this simple sequence.

P(A and B and C and D and etc.) = P(A)* P(B)*P(C)*P(D) …

P(A and B and C and D and etc.) = $.9^{13}$

P(A and B and C and D and etc.) = .2542

If each part had only a 90% probability of working properly, the overall probability of completing this simple maneuver would be only about one in four. We tend to take the reliability of our autos for granted, but what astonishing reliabilities must be built into each of their component parts for them to give us the overall reliability they do!

We have yet to illustrate the use of the special rule of multiplication.

P(A and B) = P(A) * P(B/A)

Suppose that you have a small cooler to take to a picnic. It holds six cans of soda. You put in three cans of orange drink and three cans of strawberry drink. What would be the probability that the first two friends to pull cans out of the cooler, without looking but just pulling them out at random, would both end up with cans of orange drink?

Obviously the probability of pulling out an orange drink as the first can would be 3/6 or .5, but what are the odds that the second friend would pull out a can of orange after the first event had occurred? There are now five cans left and two them are orange. So the probability of an orange on the second grab is 2/5 or .4.

P(A and B) = P(A) * P(B/A)

P(A and B) = .5 * .4

P(A and B) = .20

We might note here that the probability that something will happen is 100 % or 1.00. For example, if there is a .05 probability that some event will occur, there is a .95 probability that event would not happen. This is called the complement rule.

We also need to look at **contingency tables**. It should become evident that they are closely associated with the special rule of multiplication.

The table below shows, for the Amalgamated Aggregates company, the numbers of company associates, by educational level, employed in various types of jobs.

Educational Level of Associates

	High School	Some College	Four Year Degree	Four Years Plus	Totals
Production Worker	120	20	10	0	
Support Worker	20	20	10	0	
Mgt. or Engineering	1	5	3	1	

The first thing to be done in working with this type of table would be to complete the totals of each column and each row. (Numbers in italics.)

Educational Level of Associates

	High School	Some College	Four Year Degree	Four Years Plus	Totals
Production Worker	120	20	10	0	*150*
Support Worker	20	20	10	0	*50*
Mgt. or Engineering	1	5	3	1	*10*
	141	*45*	*23*	*1*	*210*

Question: If an associate has a four year degree what is the probability he or she will be a production worker? Answer: There are 23 individuals with four year degrees. Ten of them are production workers. The probability is 10/23 or .4348 or approximately .44.

Question: What is the probability that an associate chosen at random will have a four year degree and be a production worker. We have already seen that the probability of being a production worker if the individual has a four year degree is .44. What is the probability of an associate having a four year degree? There are 23 four degrees in a work force of 210 people, so that probability is 23/210 or .1095 or approximately .11.

We can use the special rule of multiplication:
P(A and B) = P(A)*P(B/A)
P(A and B) = .1095 * .4348
P(A and B) = .0476

Alternately we can read this probability directly off the table. Out of 210 associates there are ten production workers with four year degrees.

10/210 = .0480 The slight difference is caused by rounding to four places in calculating the event A and event B probabilities to use the formula.

While we are discussing probability, let us reproduce here a graph shown in chapter one, a diagram of a normal distribution.

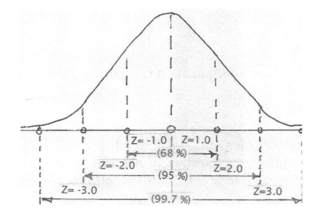

It is a characteristic of the normal distribution that approximately thirty four percent of the variables in the distribution will be located between the mean and one standard deviation to the right (or to the left, since the normal distribution is symmetrical around the mean). If you dipped into this collection of variables, what would be the probability of selecting at random a variable from either one of these areas? Approximately .34. What would be the probability of selecting at random a variable lying between minus one standard deviation and plus one standard deviation? Approximately .68 or 68 %.

We see also from the diagram that approximately 95 % of all the variables in a normal distribution are located between a point two standard deviations to the left of the mean and a point two standard deviations to the right of the mean.

Something else we will learn about the normal distribution is that five percent of the variables are located to the right of Z= plus 1.645 (measuring to the right from the mean). This means that the probability that a variable chosen at random from this distribution will be from that region (.05).

What Have You Learned and What Comes Next?

You have learned to apply the rules of probability and worked with the concept using both formulas and contingency tables. You have been introduced to the idea that characteristics of any normal distribution permit us to know the probabilities of a value in that distribution occurring close to or far away from the mean. This chapter brings us closer to the use of probabilities in hypothesis testing so problems are provided below if you would like to use them to become more familiar with the concepts covered in this chapter.

The next chapter will briefly survey the concepts of combinations, permutations, and arrangement. Although these are not directly used in hypothesis testing, a basic understanding of these concepts appears sufficiently useful to include them here.

*****

Practice Problems for this chapter.

3-1. The general special rule of multiplication is used when events are

dependent   b. independent

3-2. The Board of Directors of General Aluminum, Inc. consists of eight men and four women. A four member search committee is to be chosen at random to conduct a nationwide search for a new company president. What is the probability that all four members of the committee will be women?

		Job Category			
		Engineer A1	Clerical A2	Production A3	Total
Male	B1	30	10	60	
Female	B2	10	20	20	
Total		--------	--------	--------	--------

3-3. What is the probability of selecting an individual at random and finding that individual to be a female?

3-4. What is the probability of selecting an individual at random and finding that that individual is a male?

3-5. What is the probability of selecting an individual at random and finding that that individual is a female production worker?

3-6. What is the probability of selecting an individual at random and finding that that individual is not a production worker?

3-7. $P(A) = .40$ $P(B/A) = .30$ What is the joint probability of A and B?

3-8. You ask three strangers about their birthdays, what is the probability that all were born on Wednesday?

3-9. You have just washed your car and hope it will not rain today or tonight. The probability of rain today is .2 and the probability of rain tonight is .1. What is the probability of no rain today and no rain tonight?

3-10. The probability of visiting Rocky Mountain National Park is .6 and the probability of visiting Yellowstone is .4. It is possible to go both places. What is the probability of visiting either?

*****

Problem solutions:

3-1. Independent

3-2. What are the odds of picking the first person at random and finding it to be a woman? P(A) = 4/12 or .3333. The second woman? P(B/A) = 3/11 or .2727. The third? P(C/A,B) = 2/10 or .2000. The fourth? P(D/A,B,C) = 1/9 or .1111.

Using the special rule of multiplication:
P(A and B and C and D) = P(A)*P(B/A)*P(C/A,B)*P(D/A,B,C)
P(A and B and C and D) = .3333*.2727*.2000*.1111
P(A and B and C and D) = .0020

3-3. What is the probability of picking an individual at random and finding that person to be a female? The first thing would be to complete the table totals:

		Engineer A1	Clerical A2	Production A3	Total
		Job Category			
Male	B1	30	10	60	*100*
Female	B2	10	20	20	*50*
Total		*40*	*30*	*80*	*150*

50/150 or .3333

3-4. What is the probability of picking a person at random and finding that person to be a male? 100/150 or .6667

3-5. What is the probability of selecting an individual at random and finding that that individual is a female production worker?

20/150 or .1333

3-6. What is the probability of selecting an individual at random and finding that that individual is not a production worker? Probability of a production worker is 80/150 or .5333.

Complement rule: 1 - .5333 = .4667

3-7. P(A) = .40 P(B/A) = .30   What is the joint probability of A and B?
P(A and B) = P(A)*P(B/A)
P(A and B) = .4 * .3
P(A and B) = .12

3-8. You ask three strangers about their birthdays, what is the probability that all were born on Wednesday? These are independent events so we can use the general rule of multiplication.

P(A and B and C) = P(A)*P(B)*P(C)

What is the probability of an individual chosen at random being born on a particular day of the week? One out of seven or .1429.

P(A and B and C) = .1429 * .1429 * .1429
P(A and B and C) = .0029

3-9. You have just washed your car and hope it will not rain today or tonight. The probability of rain today is .2 and the probability of rain tonight is .1. What is the probability of no rain today and no rain tonight? These would normally be considered independent events so we can use the general rule of multiplication. If the probability of rain today is .2, the probability that it will not rain is .8 (event A). (The complement rule.) If the probability of rain tonight is .1 then the probability that it will not rain is .9 (event B).

P(A) = .8
P(B) = .9
P(A and B) = P(A)*P(B)
P(A and B) = .8*.9
P(A and B) = .72

3-10. The probability of visiting Rocky Mountain National Park is .6 and the probability of visiting Yellowstone is .4. It is possible to go both places. What is the probability of visiting either? The problem tells us that these are not mutually exclusive events so we would use the general rule of addition.

P(A or B) = P(A) + P(B) – P(A and B)

We will consider a visit to Rocky Mountain National Park as event A and a visit to Yellowstone as event B. However the problem gives P(A) = .6 and P(B) as .4 but does not provide the joint probability. If we consider these to be independent events, which seems logical, we can use the general rule of multiplication to calculate that joint probability.

P(A and B) = P(A) * P(B)
P(A and B) = .6 * .4
P(A and B) = .24

P(A or B) = P(A) + P(B) – P(A and B)

P(A or B) = .6 + .4 – .24

P(A or B) = .76

Chapter 3

Combinations, Permutations, and the Multiplication Formula

What this chapter will do for you.

The preceding chapter discussed calculation of the probability of various events taking place. Statistics can also concern itself with calculating the number of various events that could take place under various circumstances. For example: Suppose an old fashioned bank safe with a dial on the door has 100 numbers on the dial. A would-be bank robber attempting to open the safe by trying all possible combinations theoretically could spend many years and fail to succeed. The number would be very, very large. This might be a good time to note that what is called a "combination for a safe" in ordinary conversation is really a permutation. What is the difference?

Combinations. The most straightforward way to define the term combination is by example. Suppose there are three birds sitting on the limb of a tree in your yard. There is a bluebird, a redbird, and a yellowbird. On the birdfeeder under the tree there are two pegs, each of which is just large enough for one bird. The bird feeder then can accommodate only two birds at a time. Two birds at a time fly down to the bird feeder to have their breakfast. Question: How many combinations of two birds can be obtained from the three kinds of birds on the tree limb? The possibilities are listed below.

- Blue bird and red bird

- Blue bird and yellow bird

- Red bird and yellow bird

The formula that will permit us to calculate the number of combinations, without listing all the possibilities is as follows:

- **C** is the number of possible combinations

- **r** is the number of items (in this case the pegs for birds) in the combination

n is the number of items from which the combination is drawn (in this case, the number of birds sitting on the tree limb)

$$_nC_r = \frac{n!}{r!(n-r)!}$$

To state this formula in words: The number of combinations of size r that can be obtained from a pool of items of size n is equal to n factorial divided by r factorial multiplied times the quantity n minus r, factorial. We will apply that formula to our bird example.

$$_3C_2 = \frac{3!}{2!(3-2)!}$$

$$_3C_2 = \frac{1*2*3}{1*2*(1)}$$

$$_3C_2 = \frac{6}{2}$$

$$_3C_2 = \underline{3}$$

Permutations. A permutation is similar to a combination, but different in one respect: The arrangement of the items matters. In our bird example, if a red bird and a blue bird shared the bird feeder, that would be one combination. It would not matter if the red bird sat on the left peg and the blue bird on the right peg on the bird feeder, or the red bird on the right peg and the blue bird on the left. It would be the same combination. But if the birds perched one way, and then switched places it would be a different permutation.

It is apparent then that if a safe is said to have a "combination" of three numbers such as 50-25-40 this is really a permutation. If the numbers are dialed in the order stated, the safe will open. If those three numbers are dialed in any other order, it will not.

To illustrate the possible bird permutations in our bird problem we will list them below.

Blue bird and red bird **and also** red bird and blue bird (switched places)

Blue bird and yellow bird **and also** yellow bird and blue bird

Red bird and yellow bird **and also** yellow bird and red bird

We now have a total of six possible permutations.

The formula that will permit us to calculate the number of permutations, without listing all the possibilities is as follows:

P is the number of possible permutations

r is the number of items (in this case the pegs for birds) in the permutations

n is the number of items from which the combination is drawn (in this case, the number of birds sitting on the tree limb)

$$_nP_r = \frac{n!}{(n-r)!}$$

You will notice that the formula is the same as the formula for combinations except that n factorial is missing from the denominator. That makes the denominator of the fraction smaller, making the total fraction larger.

$$_3P_2 = \frac{3!}{(3-2)!}$$

$$_3P_2 = \frac{1*2*3}{(1)}$$

$$_3P_2 = \underline{6}$$

Multiplication arrangements. Another topic we need to consider in this section is the number of arrangements that we can put together using different items, when we have choices as to which specific items to use. Confusing? Once again, the clearest definition is furnished by example.

Suppose we are building automobiles. All parts of the car are the same except for three – the engine, the transmission, and the color. We can install either a four cylinder engine or a V-6, put in either an automatic transmission or a five speed manual transmission, and paint the car either red or blue. The diagram below illustrates the fact that we could produce eight "different" cars.

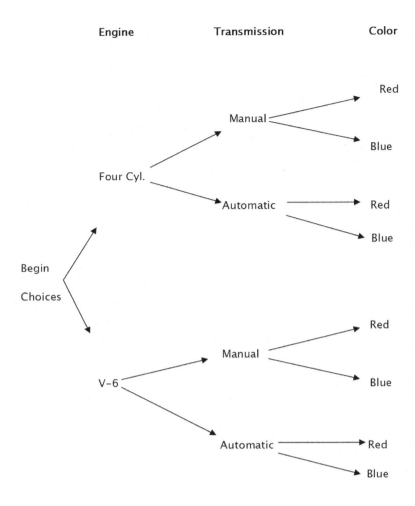

Without constructing a diagram such as the one above we can easily calculate the number of possible arrangements that could be produced by using the multiplication formula.

    Engines   Transmissions   Colors

      2   *   2   *   2   = 8 possible arrangements

Could this be important in a commercial endeavor? Certainly it could. If a business decision is made to offer greater variety, in a retail establishment for example, problems of space and logistics become more complicated. In June 2009 for example the Wall Street Journal featured a news story about retailing which indicated that some retail chains, in response to the recession and a need to cut costs, had moved from an attempt to accommodate all tastes (a policy in the former period of economic prosperity) to a reduction of varieties, sizes, smells, and flavors - to simplify operations.

Automobile production furnishes another example of how the offering of many possible "products" can complicate operations. To illustrate, the author, some years back, needed to replace his car and felt that he had finally reached the point where he could buy a new model by making payments over three years (not necessarily the best course of action – but that is another story). He purchased a U.S. brand. At that time, quality of American-made automobiles was not good by today's standards. The manufacturer offered several body styles: Four door sedan, two door sedan, two door "hardtop", convertible, and station wagon. The manufacturer also offered three transmissions, at least four different options for radios, different trim options, upholstery options, and innumerable color options. The particular items that made up each possible finished product had to be shipped and scheduled and fed into the correct assembly line at just the right point in time. Complicated? Of course. Consider a hypothetical example.

Engines Transmissions Radios Trim lines Upholstery Color Combinations

4 * 3 * 4 * 3 * 3 * 20 = 8,640 different arrangements.

By contrast, the next new car he purchased was a Japanese make. It came in one body style, with two engine choices, two transmission choices, and four possible colors. There were only sixteen possible different arrangements. The quality of the vehicle was excellent. Was there a connection? It is possible that fewer different arrangements made it more feasible for the Japanese company's quality philosophy to make itself felt.

What have you learned and what comes next?

You know the difference between combinations and permutations and can now calculate the number of combination and permutations that can exist under various conditions (size of combination or permutation and number of variables being drawn from to create them) and possess an understanding of the number of arrangement possible and how these increase with options for each element of the arrangement. Although not directly relevant to the process of hypothesis testing it is worthwhile background. The next chapter will briefly discuss methods for presentation of results that might be useful in the preparation of reports.

Problems for Combinations and Permutations.

3.1 How many combinations are possible selecting groups of 2 from a population of 5?

3.2 How many permutations are possible selecting groups of 2 from a population of 5?

3.3 If Aunt Harriet has three hats that are each red, white, and blue, Three dresses that are each red, white, and blue, and three jackets that are each red, white, and blue – How many different "outfits" can she put together to wear?

*****

Solutions

3.1
$$nCr = \frac{n!}{r!(n-r)!}$$

$$nCr = \frac{5!}{2!(5-2)!}$$

$$nCr = \frac{1*2*3*4*5}{1*2(1*2*3)}$$

$$nCr = \frac{120}{12}$$

$$nCr = \underline{10}$$

3.2
$$nPr = \frac{n!}{(n-r)!}$$

$$nPr = \frac{5!}{(5-2)!}$$

$$nPr = \frac{1*2*3*4*5}{1*2*3}$$

$$nPr = \frac{120}{6}$$

$$nPr = \underline{20}$$

3.3 Three hats times three dresses times three coats = 27 different outfits.

Chapter 4

A Few Ways to Present Data

What this chapter will do for you.

The old saying that a picture is worth a thousand words has been attributed to Napoleon Bonaparte and also to various unnamed oriental philosophers. It may or not be true, but there is little doubt that graphs and illustrations that picture a situation or relationships of data tend to make understanding of a report easier and can add to its aura of professionalism. This chapter, while quite brief, may suggest ideas for presenting data effectively.

Line Charts

Newspapers and investment newsletters often use **line charts** to show the relationship of two or more variables over time. For example, if an investment advisor was correct in advising the purchase of a particular security in January 2014 and it outperformed the broader market over the next year, a chart illustrating that fact might look something like the one below.

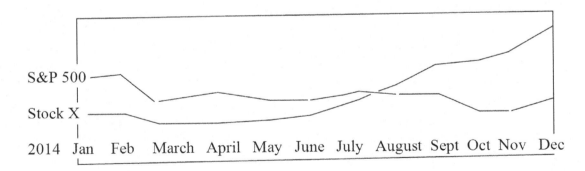

Pie Chart

A **pie chart** often seems a natural way to present to a reader how something is divided up. For example your state income tax instruction booklet, if you live in a state that has an income tax, might use one pie chart show the various elements of total tax collections and another to show where funds go.

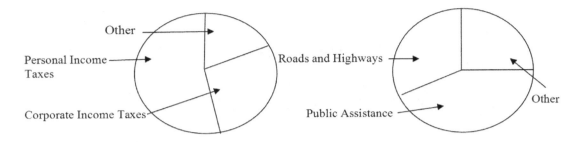

Bar Graphs.

A bar graph is normally used to show categorical data – that is, how many variables are in each of various categories.. Categories are set up and numbers in each category are shown. Convention usually has a bar graph "standing on end" as shown immediately below, however they could be shown horizontally.

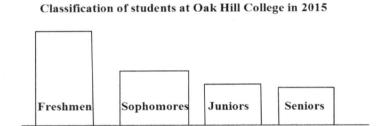

The "bars" of a bar chart are usually shown separated by spaces, however these classes can be shown adjacent to each other, with no space in between as in the illustration below.

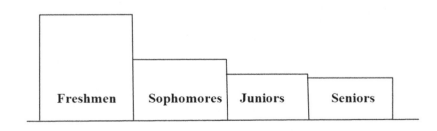

Either chart can show at a glance that the attrition rate for First Year Students at Oak Hill College was almost fifty percent in the previous year, a fact you might wish to emphasize. In a bar graph, bars are typically arranged in order of decreasing height. The bars in a histogram are conventionally displayed in the order that the classes occur. In the bar graph above, college class categories are presented left to right, largest to smallest. However this general rule could be violated and it could have been constructed with seniors presented on the left and subsequent categories on the right.

Histograms

A histogram is normally used for continuous data or grouped numerical data. Let us look at a possible example. Each day at the close of practice a high school basketball player shoots ten foul shots or free throws. Sometimes he hits most of them; sometimes he hits only a few. Usually he shoots close to his average.

Let us assume that these are his results over the last twenty practices:

2,3,4,4,5,5,5,5,6,6,6,6,6,7,7,7,7,8,8,9,10. We could present these numbers in this manner. In effect we are "piling" these variables on top of like variables.

```
                        6
                  5     6     7
                  5     6     7
            4     5     6     7     8
      3     4     5     6     7     8     9
2     3     4     5     6     7     8     9     10
────────────────────────────────────────────────────
```

(Note: It is not possible to make a partial basket. It is either a whole basket or nothing. Hence the name, a "discrete" probability distribution. Note: These are tabulations of results from an experiment. Each outcome is different. It is a distribution of results. The results form a **Binomial Distribution** showing the results when something is counted a number of times.)

Now let's add the bars to make a histogram. (Normally numbers would not be shown except that each bar would be labeled.)

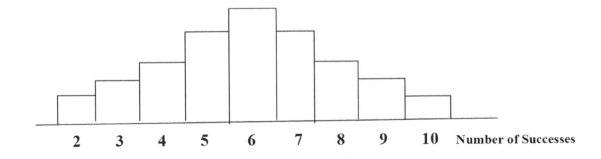

The bars in bar graphs are usually separated. Sometimes you see bar graphs with no spaces between the bars, but references tell us that histograms are never drawn with spaces between the bars. If data are divided into a frequency distribution, the frequency of each class is nominal data, however, the classes show continuous data grouped into classes, so it seems logical that the categories (or "bars") should be shown adjacent to each other.

Do you recall the frequency distribution in Chapter One? As another illustration it might be useful to construct a histogram for that data. The class limits were 3.5 up to 4.5, 4.5 up to 5.5, 5.5 up to 6.5, 6.5 up to 7.5, and 7.5 up to 8.5. The respective frequencies were 1, 2, 3, 2, and 1.

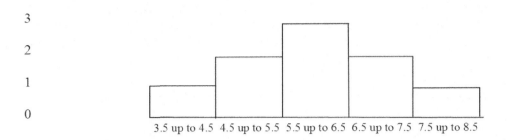

Stem and Leaf Charts

The writer has never seen a **stem and leaf chart** used in a newspaper, magazine, or scholarly article. Nevertheless, they are covered in most texts so a brief mention seems appropriate here.

Suppose you have the following measurements – the ages of the individuals playing on the swings and other apparatus at a playground on a Sunday afternoon: 8,9,10,10, 12, 13, 13,14,15,15, 17, 18, 21,22,23,23,23, 25 27,35, 37, 38, 38,40, 44, 44, 56, 57. If you wanted to display this information in a stem and leaf chart you would write down the first number of each measurement down the left side of the paper, and then add the last digit of each measurement as a "leaf" on the same line as the appropriate "stem".

5	6,7
4	0,4,4
3	5,7,8,8
2	1,2,3,3,3,5,7
1	0,0,2,3,3,4,5,5,7,8
0	8,9

Box Plots

A **box plot** shows where the middle half of a group of numbers is and also, by means of "whiskers" how far away the smallest and largest variables or numbers are from this middle half. We will use the numbers from the previous page: 8,9,10,10, 12, 13, 13,14,15,15, 17, 18, 21,22,23,23,23, 25 27,35, 37, 38, 38,40, 44, 44, 56, 57.

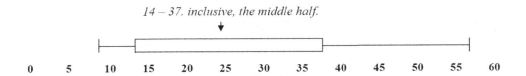

What have you learned and what comes next?

Chapter Four has presented a few common ideas for presenting data. In Chapter Five our study of hypothesis testing begins in earnest. That chapter will attempt to explain the normal distribution and the standard normal distribution in clear, down to earth language. Understanding of the normal distribution and the standard normal distribution will be fundamental to a quick and easy understanding of testing hypotheses using large samples. Further, understanding of these procedures will smooth the way for understanding the use of other distributions covered in later chapters (the t distribution, the F distribution, and Chi Square).

Chapter 5

The Normal Distribution and the Standard Normal Distribution

What this chapter will do for you.

This introduction will help you understand the normal distribution and the standard normal distribution. The understanding you develop will give you a foundation for the study of hypothesis testing using not only the standard normal distribution (the Z distribution) but other distributions as well.

The Normal Distribution.

A normal distribution can be thought of as a pile of variables. Many middle sized variables (X's) are piled in the center. Quite a few variables (X's) larger than middle sized are piled to the right, and quite a few smaller than middle sized (X's) are piled to the left. A few very large variables (or X's) are scattered out to the right, and a few very small variables (X's) as scattered out to the left.

```
                        x
                    xxxxxxx
                   xxxxxxxxxxx
                  xxxxxxxxxxxxxx
                 xxxxxxxxxxxxxxxxx
                xxxxxxxxxxxxxxxxxxxx
               xxxxxxxxxxxxxxxxxxxxxxxx
              xxxxxxxxxxxxxxxxxxxxxxxxxxx
             xxxxxxxxxxxxxxxxxxxxxxxxxxxxxx
            xxxxxxxxxxxxxxxxxxxxxxxxxxxxxxxxx
           xxxxxxxxxxxxxxxxxxxxxxxxxxxxxxxxxxxxxx
```

Although the illustration may make it appear so, these X's are not necessarily "discrete values" – numbers that can assume only certain values and that have gaps between them (often whole number). Imagine variables (X's) that can assume <u>any</u> value (such as 11.5, 12.7777713, 99. 999999999999 etc). Such numbers make up the continuous distribution we will be working with in the examples in this chapter.

The right and left sides of the distribution are symmetrical and are separated by the value that is the arithmetic mean. This value is also the median of distribution (the middle value – just as many values are larger as there are smaller). This value is also the mode (the value that occurs most often, as shown by the peak, or highest point, of the curve). Theoretically, the normal curve does not "sit on" the baseline; it "floats", that is, in theory the right and left tails extend an infinite distance to the right and to the left but never quite reach the baseline.

Any normal distribution can be described by two things: Its mean (or average): that is, where its middle value is, and its standard deviation, which measures how wide, or spread out, the distribution is.

What is a "standard deviation"? We could almost think of it as one sixth of the way across a normal distribution. That is not quite correct because a normal distribution does not have a left or right side – it is asymptotic, that is, as noted above, the line describing it "floats" – never quite touching the zero line.

However, most of the variables in the distribution (99.7%, or all but three tenths of a percent of the variables) would lie between two imaginary walls we could set up, one three standard deviations to the right of the mean, and the other three standard deviations to the left of the mean.

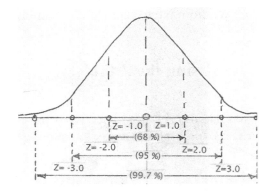

You will notice that on this illustration the term "standard deviation" is not used. The letter Z is used instead. This was done for two reasons: (1) There was not space to write "One Standard Deviation", "Two Standard Deviations", and so on. (2) The Z number is used when work is carried out with the Standard Normal Distribution. In the standard normal distribution, a distance of one standard deviation from the mean to some variable, X, is a distance of Z = 1.0.

The Standard Normal Distribution

The "Standard Normal Distribution" is commonly used in hypothesis testing with large samples. What is it? It is simply a normal distribution, lifted up off the paper and set down on top of the standard Cartesian coordinates that you are probably familiar with. In these Cartesian coordinates positive X values are measured out from the origin to the right along the horizontal axis. Negative X values are measured out from the origin to the left. Positive values of Y are measured up from the origin, and negative values down. In the standard normal distribution, negative values of Y are not used; in fact they would be meaningless. But positive values of Y stand for the frequency of each variable in the distribution. That is, the height of the curve at any point shows graphically how many variables there are of that size. For example, the curve reaches a peak at the value of the mean, because that value is also the mode – the value that occurs most frequently in the distribution. This value is also the median of the distribution. Half of the values in the distribution are larger, and half smaller. That is, visually, fifty percent of all the variables are located to the right of the mean, and half to the left.

In the standard normal distribution, distances to the right and left are measured in multiples of "Z". What does that mean? Z is just a convenient abbreviation for standard deviation. If a variable is one standard deviation to the right of the mean, Z = 1.0. If a variable is two standard deviations to the right of the mean Z = 2.0. If a variable is one and a half standard deviations to the right of the mean, Z = 1.5.

And so on. Since negative distances are measured to the left of the mean, if a variable, for example, is one and one half standard deviations to the left of the mean, Z = - 1.5.

From the illustration above, we know approximately what proportion of all the variables in a distribution are located one, two, or three standard deviations from the mean. We know that from the mean out to the right (or to the left) one standard deviation (or a distance of Z = 1 or Z= - 1) the area under the curve (the same as the proportion of all the variables in the distribution) included about 34 % of all the variables in the distribution. But what if we measured out to a point one and one half standard deviations from the mean: could we then find what proportion of all the variables in the distribution would be included between the mean and that point? The answer is yes. To find out what proportion of the variables would be under the curve between the mean and Z = 1.5 (or any other distance not an even multiple of the standard deviation) we would have to use what is sometimes simply called the " Z table. "

Using the Z table.

We can select a value for Z and use the Z table to see what proportion of all the values will lie between the mean (whatever it is) and a point that is 1.5 standard deviations away from that mean.

Let's use that value of Z=1.5 suggested earlier to determine what proportion of the variables will be between the mean and Z=1.50.

The value through the first decimal of 1.50 (1.5) is in the leftmost column on the table. The second decimal in the Z value (in this case zero) is found in the column labeled at the top "00".

Notice that all the columns are headed zero -something. The second number gives the second decimal place in the Z value (0, 1, 2, etc.) In this case the second digit is understood to be zero. (Our Z value is 1.50.)

Using four places after the decimal (which we will commonly do) we find the proportion of all the variables between the mean and a value that is 1.5 standard deviations away from the mean is .4332 or 43.32%.

Area under the Normal Curve from 0 to X

X	0.00	0.01	0.02	0.03	0.04	0.05	0.06	0.07	0.08	0.09
0.0	0.00000	0.00399	0.00798	0.01197	0.01595	0.01994	0.02392	0.02790	0.03188	0.03586
0.1	0.03983	0.04380	0.04776	0.05172	0.05567	0.05962	0.06356	0.06749	0.07142	0.07535
0.2	0.07926	0.08317	0.08706	0.09095	0.09483	0.09871	0.10257	0.10642	0.11026	0.11409
0.3	0.11791	0.12172	0.12552	0.12930	0.13307	0.13683	0.14058	0.14431	0.14803	0.15173
0.4	0.15542	0.15910	0.16276	0.16640	0.17003	0.17364	0.17724	0.18082	0.18439	0.18793
0.5	0.19146	0.19497	0.19847	0.20194	0.20540	0.20884	0.21226	0.21566	0.21904	0.22240
0.6	0.22575	0.22907	0.23237	0.23565	0.23891	0.24215	0.24537	0.24857	0.25175	0.25490
0.7	0.25804	0.26115	0.26424	0.26730	0.27035	0.27337	0.27637	0.27935	0.28230	0.28524
0.8	0.28814	0.29103	0.29389	0.29673	0.29955	0.30234	0.30511	0.30785	0.31057	0.31327
0.9	0.31594	0.31859	0.32121	0.32381	0.32639	0.32894	0.33147	0.33398	0.33646	0.33891
1.0	0.34134	0.34375	0.34614	0.34849	0.35083	0.35314	0.35543	0.35769	0.35993	0.36214
1.1	0.36433	0.36650	0.36864	0.37076	0.37286	0.37493	0.37698	0.37900	0.38100	0.38298
1.2	0.38493	0.38686	0.38877	0.39065	0.39251	0.39435	0.39617	0.39796	0.39973	0.40147
1.3	0.40320	0.40490	0.40658	0.40824	0.40988	0.41149	0.41308	0.41466	0.41621	0.41774
1.4	0.41924	0.42073	0.42220	0.42364	0.42507	0.42647	0.42785	0.42922	0.43056	0.43189
1.5	0.43319	0.43448	0.43574	0.43699	0.43822	0.43943	0.44062	0.44179	0.44295	0.44408

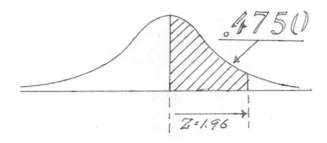

The illustration above shows a commonly used Z value, and the corresponding "area under the curve" (the proportion of all the values in the distribution) between the mean and a point that distance from the mean. This is a commonly used Z value because two and one half percent of all the variables are left out in the tail of the distribution, beyond the point where Z = 1.96.

We showed earlier that if we measured out from the mean each way one standard deviation, and put up a fence, we would enclose about 68% of all the values in the distribution. Half of 68% would be 34%. What proportion of the variables, exactly, would be between the mean and 1.0 Z?

As shown on the next page, we find this value on the Z table by reading down the left side of the table to 1.0 and then following the row across to the first column, the "00" column. When we do this we find the proportion of the values within the span (sometimes called "the area under the curve") to be .3413 or slightly more than 34 percent.

Area under the Normal Curve from 0 to X

x	0.00	0.01	0.02	0.03	0.04	0.05	0.06	0.07	0.08	0.09
0.0	0.00000	0.00399	0.00798	0.01197	0.01595	0.01994	0.02392	0.02790	0.03188	0.03586
0.1	0.03983	0.04380	0.04776	0.05172	0.05567	0.05962	0.06356	0.06749	0.07142	0.07535
0.2	0.07926	0.08317	0.08706	0.09095	0.09483	0.09871	0.10257	0.10642	0.11026	0.11409
0.3	0.11791	0.12172	0.12552	0.12930	0.13307	0.13683	0.14058	0.14431	0.14803	0.15173
0.4	0.15542	0.15910	0.16276	0.16640	0.17003	0.17364	0.17724	0.18082	0.18439	0.18793
0.5	0.19146	0.19497	0.19847	0.20194	0.20540	0.20884	0.21226	0.21566	0.21904	0.22240
0.6	0.22575	0.22907	0.23237	0.23565	0.23891	0.24215	0.24537	0.24857	0.25175	0.25490
0.7	0.25804	0.26115	0.26424	0.26730	0.27035	0.27337	0.27637	0.27935	0.28230	0.28524
0.8	0.28814	0.29103	0.29389	0.29673	0.29955	0.30234	0.30511	0.30785	0.31057	0.31327
0.9	0.31594	0.31859	0.32121	0.32381	0.32639	0.32894	0.33147	0.33398	0.33646	0.33891
1.0	0.34134	0.34375	0.34614	0.34849	0.35083	0.35314	0.35543	0.35769	0.35993	0.36214
1.1	0.36433	0.36650	0.36864	0.37076	0.37286	0.37493	0.37698	0.37900	0.38100	0.38298
1.2	0.38493	0.38686	0.38877	0.39065	0.39251	0.39435	0.39617	0.39796	0.39973	0.40147
1.3	0.40320	0.40490	0.40658	0.40824	0.40988	0.41149	0.41308	0.41466	0.41621	0.41774
1.4	0.41924	0.42073	0.42220	0.42364	0.42507	0.42647	0.42785	0.42922	0.43056	0.43189
1.5	0.43319	0.43448	0.43574	0.43699	0.43822	0.43943	0.44062	0.44179	0.44295	0.44408
1.6	0.44520	0.44630	0.44738	0.44845	0.44950	0.45053	0.45154	0.45254	0.45352	0.45449
1.7	0.45543	0.45637	0.45728	0.45818	0.45907	0.45994	0.46080	0.46164	0.46246	0.46327
1.8	0.46407	0.46485	0.46562	0.46638	0.46712	0.46784	0.46856	0.46926	0.46995	0.47062
1.9	0.47128	0.47193	0.47257	0.47320	0.47381	0.47441	0.47500	0.47558	0.47615	0.47670

Let's work a problem with the standard normal distribution. Suppose a normal distribution has a mean of 12.2 and a standard deviation of 2.5.

What is the Z distance associated with a value of 14.3?

How do we figure out what the Z value is? If the standard deviation were an even number (such as 2) it would be intuitive to calculate that one and one half standard deviations (for example) would be 3. But for something less straightforward, such as a standard deviation of 2.5, we would need to use the standard Z formula. Use of the Z formula, as illustrated on the next page, will give us the needed Z value. When we do this, we are – in textbook terminology – converting a normal distribution to the <u>standard</u> normal distribution. The mean of the distribution will be represented by the lower case Greek letter Mu (μ). The standard deviation of the population by the lower case Greek letter Sigma (σ).

$$Z = \frac{X - \mu}{\sigma}$$

$$Z = \frac{14.3 - 12.2}{2.5}$$

$$Z = \frac{2.1}{2.5} = 0.84$$

Having found that the Z value is 0.84 we then can go to the Z table to find the proportion of all the variables in the distribution (also sometimes called "the area under the curve". The use of a value of Z = .84 is shown by the arrows.

```
              Area under the Normal Curve from 0 to X

  X       0.00     0.01     0.02     0.03     0.04     0.05     0.06     0.07     0.08     0.09

  0.0   0.00000  0.00399  0.00798  0.01197  0.01595  0.01994  0.02392  0.02790  0.03188  0.03586
  0.1   0.03983  0.04380  0.04776  0.05172  0.05567  0.05962  0.06356  0.06749  0.07142  0.07535
  0.2   0.07926  0.08317  0.08706  0.09095  0.09483  0.09871  0.10257  0.10642  0.11026  0.11409
  0.3   0.11791  0.12172  0.12552  0.12930  0.13307  0.13683  0.14058  0.14431  0.14803  0.15173
  0.4   0.15542  0.15910  0.16276  0.16640  0.17003  0.17364  0.17724  0.18082  0.18439  0.18793
  0.5   0.19146  0.19497  0.19847  0.20194  0.20540  0.20884  0.21226  0.21566  0.21904  0.22240
  0.6   0.22575  0.22907  0.23237  0.23565  0.23891  0.24215  0.24537  0.24857  0.25175  0.25490
  0.7   0.25804  0.26115  0.26424  0.26730  0.27035  0.27337  0.27637  0.27935  0.28230  0.28524
→ 0.8   0.28814  0.29103  0.29389  0.29673  0.29955  0.30234  0.30511  0.30785  0.31057  0.31327
  0.9   0.31594  0.31859  0.32121  0.32381  0.32639  0.32894  0.33147  0.33398  0.33646  0.33891
  1.0   0.34134  0.34375  0.34614  0.34849  0.35083  0.35314  0.35543  0.35769  0.35993  0.36214
```

At a Z value of 0.84 the value in the Z table is .2995 or 29.95%.

Note: If we were measuring to the left of the mean (to 10.1) the Z value would be minus .84, and the proportion of all the values between the mean and 10.1 (the area under the curve) would be the same, that is, .2995 or 29.95 percent.

Why are we learning about the standard normal distribution? Because, when we start hypothesis testing, we will be working with a normal distribution called the distribution of sample means (some books call it the sampling distribution of the means) which itself is a normal distribution.

More practice with the standard normal distribution.

For a practice problem, we will assume that a normal distribution has a mean of 80 and a standard deviation of 14.

Could we find the probability of drawing a variable at random from this distribution that would be between 75 and 90?

The probability of drawing a particular value at random out of any distribution and finding it to be between two particular values is equal to the proportion of all the values in the distribution that will be between these two values.

The area under the curve between two points is the same as the proportion of all values in the distribution that lie between these two points. For example, to say that the 'area under the curve' above the mean is 50% of the total area under the normal curve is the same as saying that the proportion of all values in a distribution above the mean is .5000 or 50%.

Another example: If about 34% of the values in a distribution are between the mean and plus one standard deviation (that is – plus one Z) then the probability of picking a value at random from that distribution and finding it to be between the mean and plus one Z (or minus one Z) would be about .34

Working with the standard normal distribution

Suppose we are working with a normal distribution that has a mean of 80 and a standard deviation of 14. Can we find the probability of drawing a variable at random from this distribution that would be between 75 and 90? We can, and now that we are familiar with the Z table, the problem is relatively simple.

Our problem has two parts.

We must find the area under the curve between 75 and 80, and also find the area under the curve between the mean of 80 and a value of 90.

A picture of this two part problem is found below.

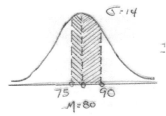

Working both parts:

$$Z = \frac{X - \mu}{\sigma} \qquad\qquad Z = \frac{X - \mu}{\sigma}$$

$$Z = \frac{75 - 80}{14} \qquad\qquad Z = \frac{90 - 80}{14}$$

$$Z = \frac{-5}{14} \qquad\qquad Z = \frac{10}{14}$$

$$Z = -.3571 \qquad\qquad Z = 0.7142$$

$$Z = -0.36 \qquad\qquad Z = 0.71$$

We can then go to the Z table to look up values of .36 and .71 to find the proportion of all the values in the distribution that will be between the mean and these two values, respectively.

A picture of the solution

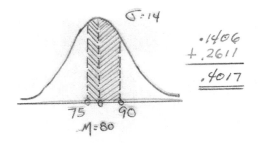

Using the same distribution of values– A slightly harder problem

What is the probability of a value between 55 and 70?
Again, we have two separate problems to work.
We need to find the area between 55 and 80.
We need to find the area between 70 and 80.

Calculations to find both areas are shown below.

Working both parts of the problem:

$$Z = \frac{X - \mu}{\sigma} \qquad\qquad Z = \frac{X - \mu}{\sigma}$$

$$Z = \frac{55 - 80}{14} \qquad\qquad Z = \frac{70 - 80}{14}$$

$$Z = \frac{-25}{14} \qquad\qquad Z = \frac{-10}{14}$$

$$Z = -1.79 \qquad\qquad Z = -0.71$$

We can then go to the Z table to look up values of -1.79 and -.71, to find the proportion of all the values in the distribution that will be between the mean and these two values, respectively.

We find in the Z table that the proportion of the variables between the mean and a point where Z = 1.79 (or between the mean and a point where Z = - 1.79) is .4633. The proportion of the variables between the mean and a point where Z = .71 (or between the mean and a point where Z = -.71) is .2022.

The difference between the two is .4633 minus .2022.

The solution is illustrated below.

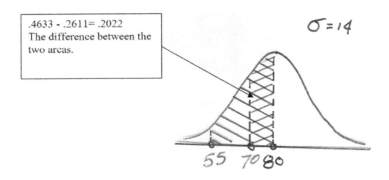

.4633 - .2611= .2022
The difference between the two areas.

$\sigma = 14$

55 70 80

Let's work a more difficult problem using the standard normal distribution: What is the value above which 80% or the variables will lie?

If fifty percent of the variables in the distribution are greater than the mean (80), then another thirty percent to the left of the mean, between the mean and the value X (which we want to calculate) would give us a total of eighty percent of the variables greater than X. This is shown in the illustration below on this page.

Fifty percent of the variables are larger than the mean.

An additional thirty percent of the variables, between the mean and X, would total 80 percent.

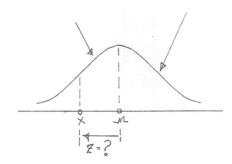

We can use the Z formula, and the Z table to find the value for X.

$$Z = \frac{X - \mu}{\sigma}$$

$$Z = \frac{X - 80}{14}$$

We have a formula with two unknowns. We can't solve that. We need to use the Z table to find a Z value to put in the formula, so we can solve for X. We go to the Z table and look in the body of the table for thirty percent. If we can do that, we can use the Z table "backwards" to find what Z value would take us that far from the mean. We cannot find exactly .3000 (thirty percent). But we can find a value that is very close: .2995. We can see that this value lies in the row identified by Z = .8 and in the column identified at the top by the number .04. This tells us that between the mean of any standard normal distribution and a point (X) a distance away of Z = 0.84 the proportion of variables under the curve will be slightly less than thirty percent.

```
          Area under the Normal Curve from 0 to X

  X       0.00    0.01    0.02    0.03    0.04    0.05    0.06    0.07    0.08    0.09

  0.0    0.00000 0.00399 0.00798 0.01197 0.01595 0.01994 0.02392 0.02790 0.03188 0.03586
  0.1    0.03983 0.04380 0.04776 0.05172 0.05567 0.05962 0.06356 0.06749 0.07142 0.07535
  0.2    0.07926 0.08317 0.08706 0.09095 0.09483 0.09871 0.10257 0.10642 0.11026 0.11409
  0.3    0.11791 0.12172 0.12552 0.12930 0.13307 0.13683 0.14058 0.14431 0.14803 0.15173
  0.4    0.15542 0.15910 0.16276 0.16640 0.17003 0.17364 0.17724 0.18082 0.18439 0.18793
  0.5    0.19146 0.19497 0.19847 0.20194 0.20540 0.20884 0.21226 0.21566 0.21904 0.22240
  0.6    0.22575 0.22907 0.23237 0.23565 0.23891 0.24215 0.24537 0.24857 0.25175 0.25490
  0.7    0.25804 0.26115 0.26424 0.26730 0.27035 0.27337 0.27637 0.27935 0.28230 0.28524
  0.8    0.28814 0.29103 0.29389 0.29673 0.29955 0.30234 0.30511 0.30785 0.31057 0.31327
```

We will substitute that value, Z = .84, in the Z formula to calculate the value of X. Because we are measuring out to the left of the mean, we must put a minus sign in front of the Z value.

$$-0.84 = \frac{X - 80}{14}$$

Multiplying both sides of the equation by 14 to get rid of the fraction:

$$14(-0.84) = X - 80$$

$$-11.76 = X - 80$$

Adding 11.76 to both sides of the equation and subtracting X from both sides of the equation puts the unknown on the left side of the equal sign and all the numbers on the right side:

$$-X = -80 + 11.76$$

Multiplying both sides of the equation by minus one will give us a positive X.

$$X = 80 - 11.76 = 68.24$$

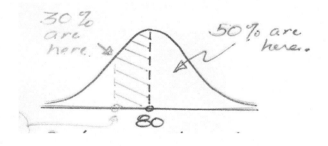

80 % of the values in the distribution will be larger than 68.24

What have you learned and what comes next?

You have learned how to work with a normal distribution by, first, using the Z formula to convert the distribution you are concerned with to the Standard Normal Distribution and, second, using the Z table. You have learned how to use the Z table directly to find the area under the curve (the proportion of all the variables in the distribution) between the mean of the distribution and a known value of X. You have also learned how to use the Z table in reverse, by determining what the area under the curve must be between the mean and an unknown value of X. Having done that, you have learned how to put that value of Z into the Z formula to calculate the value of X. In the next chapter you will learn how to apply this knowledge to perform a hypothesis test.

Chapter Questions

5-1. The text says that any normal distribution can be described by two things. What are they?

5-2. The text says that the standard deviation of a normal distribution is approximately one sixth of the distance across a normal distribution. Explain that statement.

5-3. The text says that the standard deviation of a normal distribution is not exactly one sixth of the distance across a normal distribution. Explain that statement.

5-4. If you were to measure out one standard deviation each way (plus and minus, that is, to the right and to the left) and put up a fence, you would enclose approximately what percentage of all the variables in the distribution?

5-5. If you were to measure out two standard deviations each way (plus and minus, that is, to the right and to the left) and put up a fence, you would enclose approximately what percentage of all the variables in the distribution?

5-6. If you were to measure out three standard deviations each way (plus and minus, that is, to the right and to the left) and put up a fence, you would enclose approximately what percentage of all the variables in the distribution?

5-7. If a point is located one standard deviation to the right of the mean, the distance to that point would be Z = what value?

5-8. If a point is located two standard deviations to the left of the mean, the distance to that point would be Z = what value?

5-9. Write the formula that is used to convert any normal distribution to the standard normal distribution.

5-10. A normal distribution has a mean of 100 and a standard deviation of ten. What is the Z value of a variable of 115?

5-11. A normal distribution has a mean of 100 and a standard deviation of ten. What is the Z value of a variable of 87?

5-12. A normal distribution has a mean of 100 and a standard deviation of ten. What proportion of all the variables in the distribution will be between the mean and 115?

5-13 A normal distribution has a mean of 100 and a standard deviation of ten. What proportion of all the variables will be between the mean and 87?

5-14. A normal distribution has a mean of 100 and a standard deviation of ten. What proportion of all the variables will be greater than 87?

5-15. A normal distribution has a mean of 100 and a standard deviation of ten. What proportion of all the variables will be smaller than 87?

5-16. A normal distribution has a mean of 100 and a standard deviation of ten. What proportion of all the variables will be between 87 and 115?

5-17. A normal distribution has a mean of 100 and a standard deviation of ten. What proportion of all the variables will be between the mean and 120?

5-18. A normal distribution has a mean of 100 and a standard deviation of ten. What proportion of all the variables will be between the 115 and 120?

5-19. A normal distribution has a mean of 100 and a standard deviation of ten. The largest ten percent of the variables will be how large or larger?

5-20. You are in charge of tryouts for a professional football team that is looking for two or three wide receivers with exceptional foot speed. Coaches feel that they can teach the other skills (catching the ball, running after the catch, etc.) needed to a select number of athletes, if they have a small pool of fast players to select from and work with. You test a number of would-be candidates (football players, track athletes, and others, and find that the average time in the 100 yard dash is 12 seconds. Times are normally distributed, with a standard deviation of .5 seconds. You are instructed to sign the fastest one percent of the candidates to a short term tryout contract. What time in the 100 yard dash would a candidate need to be awarded a contract?

Area under the Normal Curve from 0 to X

Area under the Normal Curve from 0 to X

x	0.00	0.01	0.02	0.03	0.04	0.05	0.06	0.07	0.08	0.09
0.0	0.00000	0.00399	0.00798	0.01197	0.01595	0.01994	0.02392	0.02790	0.03188	0.03586
0.1	0.03983	0.04380	0.04776	0.05172	0.05567	0.05962	0.06356	0.06749	0.07142	0.07535
0.2	0.07926	0.08317	0.08706	0.09095	0.09483	0.09871	0.10257	0.10642	0.11026	0.11409
0.3	0.11791	0.12172	0.12552	0.12930	0.13307	0.13683	0.14058	0.14431	0.14803	0.15173
0.4	0.15542	0.15910	0.16276	0.16640	0.17003	0.17364	0.17724	0.18082	0.18439	0.18793
0.5	0.19146	0.19497	0.19847	0.20194	0.20540	0.20884	0.21226	0.21566	0.21904	0.22240
0.6	0.22575	0.22907	0.23237	0.23565	0.23891	0.24215	0.24537	0.24857	0.25175	0.25490
0.7	0.25804	0.26115	0.26424	0.26730	0.27035	0.27337	0.27637	0.27935	0.28230	0.28524
0.8	0.28814	0.29103	0.29389	0.29673	0.29955	0.30234	0.30511	0.30785	0.31057	0.31327
0.9	0.31594	0.31859	0.32121	0.32381	0.32639	0.32894	0.33147	0.33398	0.33646	0.33891
1.0	0.34134	0.34375	0.34614	0.34849	0.35083	0.35314	0.35543	0.35769	0.35993	0.36214
1.1	0.36433	0.36650	0.36864	0.37076	0.37286	0.37493	0.37698	0.37900	0.38100	0.38298
1.2	0.38493	0.38686	0.38877	0.39065	0.39251	0.39435	0.39617	0.39796	0.39973	0.40147
1.3	0.40320	0.40490	0.40658	0.40824	0.40988	0.41149	0.41308	0.41466	0.41621	0.41774
1.4	0.41924	0.42073	0.42220	0.42364	0.42507	0.42647	0.42785	0.42922	0.43056	0.43189
1.5	0.43319	0.43448	0.43574	0.43699	0.43822	0.43943	0.44062	0.44179	0.44295	0.44408
1.6	0.44520	0.44630	0.44738	0.44845	0.44950	0.45053	0.45154	0.45254	0.45352	0.45449
1.7	0.45543	0.45637	0.45728	0.45818	0.45907	0.45994	0.46080	0.46164	0.46246	0.46327
1.8	0.46407	0.46485	0.46562	0.46638	0.46712	0.46784	0.46856	0.46926	0.46995	0.47062
1.9	0.47128	0.47193	0.47257	0.47320	0.47381	0.47441	0.47500	0.47558	0.47615	0.47670
2.0	0.47725	0.47778	0.47831	0.47882	0.47932	0.47982	0.48030	0.48077	0.48124	0.48169
2.1	0.48214	0.48257	0.48300	0.48341	0.48382	0.48422	0.48461	0.48500	0.48537	0.48574
2.2	0.48610	0.48645	0.48679	0.48713	0.48745	0.48778	0.48809	0.48840	0.48870	0.48899
2.3	0.48928	0.48956	0.48983	0.49010	0.49036	0.49061	0.49086	0.49111	0.49134	0.49158
2.4	0.49180	0.49202	0.49224	0.49245	0.49266	0.49286	0.49305	0.49324	0.49343	0.49361
2.5	0.49379	0.49396	0.49413	0.49430	0.49446	0.49461	0.49477	0.49492	0.49506	0.49520
2.6	0.49534	0.49547	0.49560	0.49573	0.49585	0.49598	0.49609	0.49621	0.49632	0.49643
2.7	0.49653	0.49664	0.49674	0.49683	0.49693	0.49702	0.49711	0.49720	0.49728	0.49736
2.8	0.49744	0.49752	0.49760	0.49767	0.49774	0.49781	0.49788	0.49795	0.49801	0.49807
2.9	0.49813	0.49819	0.49825	0.49831	0.49836	0.49841	0.49846	0.49851	0.49856	0.49861
3.0	0.49865	0.49869	0.49874	0.49878	0.49882	0.49886	0.49889	0.49893	0.49896	0.49900
3.1	0.49903	0.49906	0.49910	0.49913	0.49916	0.49918	0.49921	0.49924	0.49926	0.49929
3.2	0.49931	0.49934	0.49936	0.49938	0.49940	0.49942	0.49944	0.49946	0.49948	0.49950
3.3	0.49952	0.49953	0.49955	0.49957	0.49958	0.49960	0.49961	0.49962	0.49964	0.49965
3.4	0.49966	0.49968	0.49969	0.49970	0.49971	0.49972	0.49973	0.49974	0.49975	0.49976
3.5	0.49977	0.49978	0.49978	0.49979	0.49980	0.49981	0.49981	0.49982	0.49983	0.49983
3.6	0.49984	0.49985	0.49985	0.49986	0.49986	0.49987	0.49987	0.49988	0.49988	0.49989
3.7	0.49989	0.49990	0.49990	0.49990	0.49991	0.49991	0.49992	0.49992	0.49992	0.49992
3.8	0.49993	0.49993	0.49993	0.49994	0.49994	0.49994	0.49994	0.49995	0.49995	0.49995
3.9	0.49995	0.49995	0.49996	0.49996	0.49996	0.49996	0.49996	0.49996	0.49997	0.49997
4.0	0.49997	0.49997	0.49997	0.49997	0.49997	0.49997	0.49998	0.49998	0.49998	0.49998

Source: Engineering Statistics Handbook (National Bureau of Standards)

http://www.itl.nist.gov/div898/handbook/eda/section3/eda3671.htm

Answers to chapter questions

5-1. The mean and the standard deviation.

5-2. If you were to start at the mean of a normal distribution and measure out to the right three standard deviations, and then go back to the mean and measure out to the left three standard deviations, you would include 99.7 percent of all the variables in the distribution.

5-3. Although if you were to measure out from the mean three standard deviations each way three standard deviations, and enclose 99.7 percent of all the variables. There would be .15 percent left in each tail of the distribution. This is because the curve is asymptotic to the baseline.

5-4. Sixty eight percent.

5-5. Ninety five percent.

5-6. Ninety nine point seven percent.

5-7. $Z = 1$

5-8. $Z = -2$

5-9. $Z = \dfrac{X - \mu}{\sigma}$

5-10. $Z = 1.5$

5-11. $Z = \dfrac{87 - 100}{10} = \dfrac{-13}{10} = -1.3$

5-12. Taking $Z = 1.5$ to the Z table: .4332

5-13. Taking $= -1.3$ to the Z table: .4032

5-14. .4032 + .5000 (to the right of the mean) = .9032

5-15. .5000 - .4032 = .0968

5-16. .4032 + .4332 = .8364

5-17. $Z = \dfrac{120 - 100}{10} =$ $Z = 2.0$ From the Z table: .4772

5-18. The area under the curve between the mean and 120 minus the area under the curve between the mean and 115: .4772 - .4332 = .044

5-19. The Z table must be used in reverse to find a Z value to insert in the Z formula.

To separate the largest ten percent of the variables in the distribution in the right tail of the distribution from the remaining forty percent that are larger than the mean, we must look in the Z table for forty percent or .4000. The closest value we can find in the body of the table is .3997. This corresponds to a Z of 1.28.

We insert that value in the Z table, and solve for X.

$$1.28 = \frac{X - 100}{10}$$

$$12.8 = X - 100$$

$$-X = -100 - 12.8$$

$$X = 100 + 12.8$$

$$X = 112.8$$

5-20. The fastest candidates will have the lowest numerical values in the distribution. Therefore we will be working with the left tail where Z values are negative. The fastest one percent will have times faster (fewer seconds) than the other 49 percent of the runners with times below the mean time of 12 seconds. Therefore we look for 49 percent (0.4900) in the body of the Z table. We find a value of 0.4901. This is the value in the table closest to 0.4900. It has a corresponding Z value of 2.33. We insert that value in the Z formula, with a negative sign.

$$-2.33 = \frac{X - 12}{.5}$$

$$X = 12 - 1.17 \text{ or } X = \text{ a dash time of 10.83 seconds.}$$

Chapter 6

Sampling Theory and Introduction to Hypothesis Testing

What this chapter will do for you.

This chapter will give you an introduction to hypothesis testing. It will explain why we do hypothesis tests using samples from a large population, instead of trying to measure every item in that population. It will describe the general form of what is called the "null hypothesis" in hypothesis tests. And it will illustrate, by using an example, that sampling theory is the foundation of these hypothesis tests. When you finish studying this chapter you should understand sampling theory: that is, that the theoretical distribution of sample means (sometimes called the sampling distribution of the means) is centered on the actual population mean and is itself a normal distribution. Based on this knowledge, you will be able to conduct a hypothesis test using a single, large sample.

What does a hypothesis test do?

It is important to remember that a hypothesis test using a sample from some population does not <u>prove</u> anything about the population from which the sample was taken. It can strongly <u>suggest</u> that something is or is not true of a population, but it does no more than that. The only way we could know, for certain, what the arithmetic mean of a population is, would be to take a *census* of the population (that is measure every variable in the population) and then calculate the arithmetic mean.

Then why do we do take samples and carry out a hypothesis test? Measuring every variable in a population is very likely to be too time-consuming, too expensive, impractical, or even impossible. Suppose we have a railroad boxcar full of two by four pieces of lumber, and we want to know what the average strength of the pieces of lumber would be, if the ends were supported and an iron weight hung in the middle. We could calculate a very accurate answer if we conducted a census and broke each two by four. But then we would have no two by four's remaining to build anything, because we would have broken every one. Obviously, using a sample to <u>estimate</u> the average breaking strength of the lumber would make sense.

It might be that the population with which we were concerned could be the values that would accumulate over a period of twenty years. We could wait and accumulate the values, and at the end of twenty years, come to some conclusion, but the opportunity to take some action that would change the values then would be twenty years in the past. We would need to take a sample now, draw some conclusion, and take action – now.

The Central Limit Theorem

There is a theorem that states that repeated samples of the same size taken from a population will have a distribution of the sample means that approximately follows the Normal distribution. This allows the analyst to evaluate a sample mean as if it came from a distribution of samples means that were distributed following the normal distribution.

Something interesting happens when we take a sample out of a population: If we calculate the arithmetic mean of the sample, it tends to be close to the population mean. It won't always be close to the population mean, but will be more often than not. Why is that? Think about what would likely happen if you flipped a coin several times. You would expect to get several tails, and several heads, because there is a fifty-fifty probability of getting either one on each flip. When we remember that in a normal distribution, fifty percent of the variables are greater than the arithmetic mean of the distribution and fifty percent are smaller, we can see that a sample will almost always have a number of values larger than the mean and a number smaller. When we average larger-than and smaller-than values, the mean is going to be somewhere in the middle – often quite close to the population mean.

This tendency for sample means to be near the mean of the population from which the sample is taken holds true even if the population distribution is not a normal distribution. Imagine a distribution shaped like a brick, or a shed with a sloping roof, or a triangle on its side (or any other shape you can imagine); in each case sample means will tend to be in the middle some place, often close to the population mean.

We now recognize that that the mean of a sample taken from a population will tend to be close to the actual population mean. A few will be far away. As the number of samples increases, and the mean of each is taken, the distribution of sample means approaches a normal distribution.

The mean of the distribution of sample means will be the same value as the population mean. This is shown below.

The distribution of sample means is a normal distribution.

This is the foundation of sampling theory.

What is the basis for testing a null hypothesis? The basis is sampling theory. Sampling theory says that if you take a sample from a population, you can calculate an average value for the variables in that sample. (The sample mean.) If you took a second sample, you could calculate its mean. If you took a third sample, you could calculate its mean. And so on.

In practice, you would take only one sample, but theoretically, a large number of samples could be taken, and their means calculated. The number would be so large that it would be extremely impractical or even humanly impossible to take them all. For example, the number of samples of size thirty that could be taken from a population of 100 items could be calculated using the combination formula. That formula would tell us that the number of combinations (samples) of size "r" that could be taken from a population of size N would be very large.

Although, in this chapter, we use a small letter "n" as the number of a sample, most texts use small letter "r" in the combination formula. In this combination formula below, " r " is the size of the combination – which corresponds to the number of samples of a particular size that could be taken out of a larger population.

$$_NC_r = \frac{N!}{r!(N-r)!}$$

$$_{100}C_{30} = \frac{100!}{30!(100-30)!}$$

$$_{100}C_{30} = 2.937 \times 10^{25} \approx 29{,}370{,}000{,}000{,}000{,}000{,}000{,}000{,}000.$$

So a large number of samples could be taken – why is this important? It is important for this reason: Sampling theory says that if <u>all possible</u> samples of a particular size are taken from a population, even if the variables in that population are not normally distributed, <u>the sample means themselves would be normally distributed</u>. If you plotted this distribution of sample means they would be much less spread out (less dispersed) than the variables in the population. (This makes sense, because in most samples there would be some variables larger than the population mean and some smaller than the population mean. When these are averaged out, the sample mean would tend to be close to the population mean.)

As illustrated below, and described earlier, the distribution of sample means would be centered on the population mean. That is, the mean of the population and the mean of the distribution of sample means would be the same. Note: In this illustration it is assumed that the population from which all these samples are taken is a normal distribution, but – an noted above - that is not critical to the operation of sampling theory.

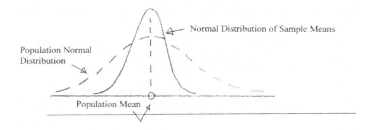

Textbooks sometimes call the distribution of sample means "the sampling distribution of the means" which may seem non-descriptive. To name it "the distribution of sample means" is to call it exactly what it is.

Since this theoretical distribution of sample means is a normal distribution, it has its own standard deviation. This standard deviation is called **The Standard Error of the Mean**. The word error does not mean that someone made a mistake, it simply means that when we sample a population and get a sample mean, there will always be some error – that is the sample mean will never be exactly what the population mean is – it will always be off to the left (smaller than the population mean) or off to the right (larger than the population mean).

An illustration of the relationship between a population mean and the distribution of the means of all possible samples of a given size that might be taken from that population.

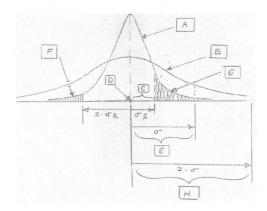

In the illustration above:

A identifies the theoretical distribution of sample means.

B identifies the population distribution.

C illustrates one standard error of the mean (that is, one standard deviation or a distance of Z = one, measured on the distribution of sample means.)

D is the population mean.

E shows one standard deviation of the population mean.

F illustrates the portion of all the sample means that will be located to the left of Z equals negative two. (The Z distance in this case applies to the theoretical distribution of sample means, not the population distribution – sometimes called "the sampling distribution of the means".)

G illustrates the portion of all sample means that will be located to the right of Z = one. (Again, this Z distance applies to the distribution of sample means.)

H shows a distance of two population standard deviations to the right of the population me

Confidence Intervals

Once you have obtained a sample from a population and determined the sample mean and the sample standard deviation, it is possible to estimate where the true mean lies of the entire population. The standard deviation and the sample size determine how far from your sample mean the true mean of the populations could be based on the normal distribution. If you find a small standard deviation with a large sample size, then you can be fairly confident that the true mean of the population will be close to the sample mean. If you find a large standard deviation with only a small sample size, then there is a lot of doubt about the location of the true mean of the population. By way of example: measuring from the sample mean to the left 1.96 times the standard error of the mean, and from the sample mean to the right 1.96 times the standard error of the mean, we set the boundaries for a 95 percent confidence interval for that mean. That is, we can be 95 percent confident that the population mean lies between those limits.

A situation calling for a hypothesis test.

Suppose that a good friend of yours is a football coach. He is not pleased with his defensive squad because of what he believes is a lack of sufficient aggressiveness. On a visit to the library last week he happened to pick up an old issue of a psychology journal and read that university professors, some years back, performed a study of the impact of color on athletes and came to the conclusion that players who wear black jerseys tend to exhibit a higher level of aggressiveness than players wearing jerseys of white, red, blue, or any other color tested.

He wonders if his team members would play more aggressively if he changed the color of their jerseys. He briefly considered getting black jerseys for his players, and keeping track of how they play over the next several seasons. (That would be a population of opportunities for his defensive squads to improve their performance.) But if he takes that much time to find out if black jerseys help, he may get fired before he can complete such a census of performance results. He asks you for help in conducting a hypothesis test to judge whether or not black jerseys might help, so he can make a decision before the season begins next September.

Taking a sample

In his intense spring training each year, squads play 36 abbreviated game-condition scrimmages with other schools in the immediate area. You suggest that, this spring, he make a change from the school's traditional green practice jerseys to black jerseys, and then measure, for the 36 scrimmages, some performance factor that can be counted, that will give a good indication of the level of aggressiveness of his defense.

Aggressiveness itself is not directly measurable. You will need something measurable to provide a numerical equivalent. Your friend, the coach, for the past several years, has conducted scrimmages similar to those planned this year. Each year, in each scrimmage, he has counted the number of defensive stops of the other team, with no positive gain in yards, because he felt that this was a good measure of aggressiveness. That number has varied somewhat over the years, but has averaged ten tackles for zero or negative gain per scrimmage. That is the population average. You suggest that he keep records for this year's sample of thirty six scrimmages, with black jerseys. He does so and finds the average number in this sample is eleven such stops per scrimmage.

It does <u>appear</u> that there has been an improvement. But this could be an accident of sampling. If black jerseys really make no difference, and he did the same thing again next year, the average of a similar sample might be ten, or nine, or eleven, or some other value. In fact, even if the black jerseys do make a difference, doing the same thing would result in different sample average values, simply because that is what happens when we take samples.

The question, then, is this: Is our one sample average (mean) different enough from the historical average (from the mean of the population) to make us confident that we can conclude that the black jerseys truly increase the aggressiveness of the coach's defensive players?

We will make the assumption here that no other variables change: the strength of squads from other schools is the same as always and there are no impacts due to weather or other factors.

(You might ask, why did we choose 36 as the number of scrimmages in our example. A sample number of thirty or more is generally considered a "large sample" and uses a particular way of determining the significance of calculated numbers. A test with a "small sample" (less than thirty) is conducted somewhat differently – as we will see in a later chapter.)

Setting up the hypothesis test.

We want to see if hypothesis testing will support the conclusion that the black jerseys have indeed resulted in an increase in the number of no-gain defensive stops. There was, as we have noted above, an apparent increase, but this could be the result of *sampling error.*

What hypothesis would we test? Many students might be inclined to state that they were going to test the hypothesis that "the number of tackles will be higher for games when the team wars black jerseys." This would not be correct. This is *not* a testable hypothesis.

The hypothesis that must be tested is called the *null hypothesis*. If the coach can find adequate reason to reject the null hypothesis, he can then accept an alternate statement called the *alternate hypothesis*. In this case the null hypothesis would be stated in words this way: "The average number of tackles for no gain in the entire population of possible black jersey games is equal to what it has been in the past. That is, it is ten no-gain tackles per game." The Greek letter mu is normally used in statistics texts for the mean of a population. The subscript, a small letter "b", indicates this is the average when the players wear black jerseys.

$H_0 : \mu_b = 10$

An alternate hypothesis is what we can believe, if we find adequate reason to reject the null hypothesis. There are three possible alternative hypotheses in this case. (1) The mean is greater than ten. (2) The mean is less than ten. (3) The mean is not equal to ten. Statement number one fits what the couch is interested in, and that will be used in this example.

$H_1 : \mu_b > 10$

Be careful setting up your null hypothesis!

Before we continue with this example, we need to say something about hypothesis testing in general, and, in particular, the form taken by the null hypothesis in various kinds of hypothesis tests. The null hypothesis will always be based on the notion that nothing has changed, there is no difference between things, or simply "nothing is going on" that implies change, difference, or cause and effect. The statements below are examples.

> ***- The population we just took a sample from is no different from what it has been like in the past.*** (The average, or mean, is unchanged.) This is the null hypothesis in the problem we are presently working, where the coach wants to determine if he should believe that the average number of tackles for no gain has actually increased because his players are wearing black jerseys. We are using a one sample "Z" test. If we had a sample of less than thirty we would use a one sample "t" test.

> ***- The mean of one population is the same as the mean of a second population.*** This could be a two sample "Z" test if both samples are 30 or larger, or a two sample "t" test if one or both samples are less than 30.

> - The mean (average) difference – for a number of pairs of variables -between one variable in each pair of variables and the second variable in the pair, when the first variable is part of one population and the second variable is part of a second population, is zero. This is called a test of paired differences, or a test using dependent samples.

> - There is no difference between the means of three, four, or more populations. They are all the same. This is called Analysis of Variance or ANOVA.

> ***- There is no association between two variables.*** In this case we would be using a Chi Square test. ("Ki", as in "kite", not "Ke", as in a "key" for a door – Square) This is a useful test when we have only nominal data available. It is done by using a contingency table, and compares observed values with expected values.

Because this distribution of sample means is a normal distribution, we can work with it using the ***Standard Normal Distribution***. For example, if we start at the mean and measure out to the right, one standard distribution, we will include approximately 34 percent of all the variables (34 percent of all the possible sample means) in the area under the normal curve between the mean and the point we have measured to. If we measure out 1.64 standard deviations from the mean, we will include 45 percent of all the sample means. You will recall from the discussion of the standard normal distribution that when we measure out one standard deviation we say that the distance involved is Z = one. When we measure out 1.645 standard deviations, $Z = 1.645$.

We must remember that a one sample test will not prove that the mean of the population in this problem is still ten, or that it is now something different. What a one sample test may let us do however, is to know whether or not we can be reasonably confident that the mean of the population is less than, greater than, or simply not equal to ten. In this case the alternate hypothesis, above, was stated as greater than ten.

Using a hypothesis test to solve the coach's problem.

Let us suppose that we find, in our sample, that the average number of tackles for no gain is greater than ten. (As you recall, the sample mean was found to be 11.) That would encourage the coach, but he does not want to make a hasty conclusion that the population mean has now increased. In our example the coach decides that he will believe that the population mean is greater than ten – that he will reject the null hypothesis and accept the alternate hypothesis – if the one sample mean he calculates when he takes one sample is so large that it would occur only five percent of the time, or less, because of an accident of sampling.

To say this again: If the one sample mean is far out in the right tail of the normal theoretical distribution of sample means, that could possibly occur if the population mean is actually still ten, but it is not likely. In this case he decides that if the sample mean is so far out in the right tail of the theoretical distribution of sample means, that it would only be that far out five percent of the time or less due to sampling accident, he will reject the null hypothesis. This would be called a "five percent level of significance."

Statisticians talk of **Type One Errors**. What is a type one error? In this case, it is the error that would occur if the population mean was still ten, after the jersey color change, but when a single sample was taken, the mean of that sample calculated out to be so large that it was in the largest five percent that would occur by accident. In this case, a five percent level of significance was chosen. This means that there is a .05 probability that the null hypothesis will be rejected when it is really correct.

Now, let us look at that distribution of sample means again. If the population mean is still ten, the center of the distribution of sample means is also ten, because the distribution of sample means is centered on the population mean. Because the distribution of sample means is a normal distribution, we know that if we measure from its mean, out to the right to the point where Z = 1.645, we will enclose 45 percent of all the variables (sample means) under the curve between the mean and that point. The remaining five percent of the sample means that are larger than the population mean will lie farther out to the right. We would say that these possible sample means lie in *the region of rejection*. That is, if our one sample mean turns out to be in this area, we will reject the null hypothesis and accept the alternate hypothesis.

This region of rejection is shown as a shaded area in the diagram below. The region of acceptance (of the null hypothesis) is the un-shaded area.

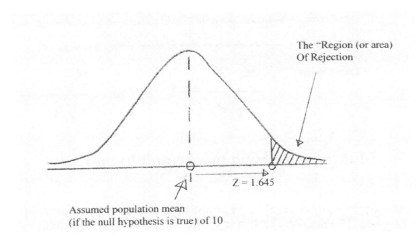

Applying the decision rule.

We have made up our null hypothesis and our alternate hypothesis. We have related them to the areas of rejection and the areas of acceptance in the diagram above. We now need to state the ***decision rule***. In this case it would be "Reject Ho (the null hypothesis) if Z is greater than 1.645." What does that mean exactly? It means that we must calculate how far the sample mean of 11 is from the former population mean of 10, in terms of "Z" units, and see whether it lies in the region of acceptance, or in the region of rejection. To do this we need to know the standard deviation of the distribution of sample means.

The standard deviation of the distribution of sample means is called the ***standard error of the mean***. The use of the word "error" does not mean that anyone has made a mistake. It simply refers to the fact that sampling always involves some amount of error. The sample mean could be almost anywhere from approximately three Z's larger than the population mean to three Z's less. (It could actually be larger or smaller than that, but the probability of this happening is very small.)

The standard error of the mean is estimated by dividing the standard deviation of the sample by the square root of the number of variables in the sample. If we knew each of the values in a sample, we could calculate the standard deviation of the sample, but in this example, we simply assume that the standard deviation of the sample (S) is two. Since the sample number (the number of variables in the sample) was 36, the standard error of the mean would be two divided by the square root of 36 – or simply two divided by six.

$$S_{\bar{x}} = 2 / \sqrt{36} \quad \text{or} \quad S_{\bar{x}} = 2/6 \quad \text{or} \quad 1/3$$

You will note that we called this an estimate of the standard error of the mean. The precise calculation of the standard error of the mean would be done by dividing the standard deviation of the population by the square root of the number of variables in the sample. But, in the real world, we would not likely know the standard deviation of the population. In the real world then, the approach would be to calculate the standard deviation of the sample and use that value to calculate, in turn, a good estimate of the standard error of the mean.

Next, we will need to perform a calculation to see how far out – in terms of units of Z - our sample mean of 11 no-gain tackles is from the assumed population mean (the former population mean) of ten. To do that we would subtract the assumed population mean from the sample mean, and divide the result of that subtraction by the standard error of the mean.

$$Z = \frac{\bar{X} - \mu}{S_{\bar{x}}} \quad \text{or} \quad Z = \frac{\bar{X} - \mu}{S / \sqrt{n}} \quad \text{or} \quad Z = \frac{11 - 10}{2 / \sqrt{36}}$$

$$Z = \frac{1}{2/6} \quad \text{or} \quad Z = 3$$

Making a decision on the null hypothesis.

Is our sample mean in the region of acceptance or rejection? Since the ***calculated value*** of the ***test statistic*** is three, and the ***critical value*** of the test statistic was 1.645, the decision rule says that our mean – which is larger than 1.645 – is in the region of rejection.

The alternate hypothesis states that the average number of tackles– following the jersey color change – is now larger than ten. By our rejection of the null hypothesis and acceptance of the alternate hypothesis, have we proven that it is greater than ten? No. We could be committing a type one error, there is a .05 or one in twenty probability of this occurring.

The approximate location of our calculated test statistic is shown in the diagram below.

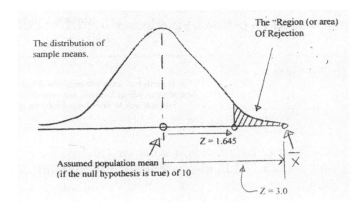

There is a .95 probability that the coach will be correct in concluding that the black jerseys result in more aggressive play by his defense, using the criterion he chose, number of defensive tackles with no gain by the opponent. If our coach can find the funds to do so, he will be ordering new game jerseys in black.

Type Two error.

You may be wondering, since there is something called a type one error, if there might be a type two error. There is. Suppose our sample mean had been smaller (perhaps ten and a half tackles). If it had been, our Z value would have calculated out to be 1.5, just within the region of acceptance. Would that have proven that the null hypothesis was a correct statement – that the population mean was actually ten? No. But we would have accepted the null hypothesis when it was quite possibly not true. That would be a type two error.

A "two tail" test using the same numbers.

Your friend the coach wanted to determine whether or not he should believe that black jerseys would raise the level of aggressiveness of his players. That was a very practical problem. Stating the alternate hypothesis as you did, looking for an increase in the number of no-gain tackles per scrimmage, made sense. That was a "one tail" statistical test. Let us examine a slightly different situation, just so we can see how a "two tail" statistical test would be different.

Suppose you know another football coach. This man believes in loud, verbal exhortation of his players, and is convinced uniforms make no difference in how hard they play. It's all the same to him whether his players wear white, black, red, green, or even pink. Someone tells him that if he changed from green to black it would make a difference in how they play. (Note that here the statement is not that they would play more effectively, but just that their level of play would be different.) He says "prove it!"

We know that a one sample hypothesis test does not prove anything (and since he has studied statistics he knows this also). But because he has studied statistics he is agreeable to conducting a one sample test of hypothesis to indicate whether or not he should change his mind.

How would two-tail hypotheses be different?

The null hypothesis would be the same.

$H_0 : \mu_b = 10$

As in the previous example, the alternate hypothesis is what we can believe is likely true, <u>if</u> we find adequate reason to reject the null hypothesis. In this case we are questioning whether or not the mean has changed, in other words, should we conclude that the mean is no longer equal to ten?

$H_1 : \mu_b \neq 10$

The calculations carried out would be the same as before, and the calculated value of the test statistic (the Z value) would be the same. What would be different? The decision rule and the location of the region of rejection would be different.

A ten percent level of significance for a two tail test.

The diagram below illustrates that, in a two tail test, the region of rejection is divided into two parts, one half in each tail of the theoretical normal distribution of sample means. To build on what we have just reviewed about areas under the curve in a normal distribution, we will use a ten percent level of significance. Now half (five percent) will be in each tail. In this case, the Z distance that separates the region of rejection from the region of acceptance is plus 1.645 and minus 1.645. (If we had used a five percent level of significance for a two tail test, there would have been two and a half percent in each tail and the critical value of Z would have been + and – 1.96. If it is not immediately apparent where these Z values came from, you need to refer back to the discussion of the standard normal distribution in Chapter Seven.)

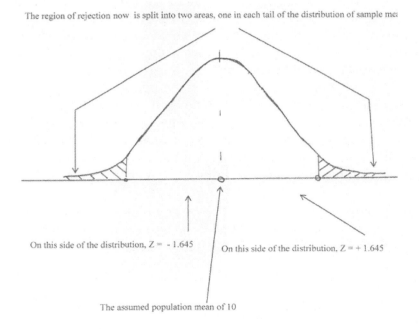

The region of rejection now is split into two areas, one in each tail of the distribution of sample means

On this side of the distribution, Z = – 1.645

On this side of the distribution, Z = + 1.645

The assumed population mean of 10

The decision rule and the decision on the null hypothesis.

The decision rule would be "Reject Ho if Z > + 1.645 or < - 1.645."

Since the calculated value of the test statistic was 3.0 the null hypothesis would be rejected, and the alternate hypothesis, that the mean was not equal to ten, would be accepted.

You might want to review this illustration of the relationship between a population mean and the distribution of the means of all possible samples of a given size that might be taken from that population.

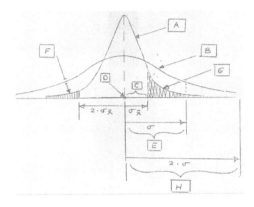

In the illustration above:

    A identifies the theoretical distribution of sample means.

    B identifies the population distribution.

    C illustrates one standard error of the mean (that is, one standard deviation or a distance of Z = one, measured on the distribution of sample means.)

    D is the population mean.

    E shows one standard deviation of the population mean.

    F illustrates the portion of all the sample means that will be located to the left of Z equals negative two. (The Z distance in this case applies to the theoretical distribution of sample means, not the population distribution – sometimes called "the sampling distribution of the means".)

    G illustrates the portion of all sample means that will be located to the right of Z = one.(Again, this Z distance applies to the distribution of sample means.)

    H shows a distance of two population standard deviations to the right of the population mean.

What you have learned, and what comes next.

You have learned the basics of hypothesis testing and how to conduct a one sample hypothesis test, using one large (n is equal to or greater than 30) sample. In the next chapter, you will learn to conduct a hypothesis test using two large samples. Now that you have learned the basics of hypothesis testing, using, as an example, the single sample test you have just been through, the other hypothesis tests you will learn will be relatively easy, because they are all based on similar logic. You will make a decision as to how the calculated value of the test statistic compares with the critical value of the test statistic. That is, does the calculated value of the test statistic place it in the region of rejection or the region of acceptance. This will be true whether you are using the "Z" statistic (from large samples), the "t" statistic (from small samples), the "F" statistic (in an analysis of variance), or the Chi-Square statistic.

Chapter Questions

6-1. Explain the difference between a sample and a census.

6-2. A marketing research company your firm has employed utilizes a carefully selected random sample to evaluate consumer preferences for various shapes of pickle bottles and tells you in a briefing that they have proven that consumers prefer square bottles. When they tell you this what would you tell them?

6-3. Gasoline has been selling for an average price of $3.00 per gallon in Ohio. Your transportation manager hears reports that the average price has gone up and decides to conduct a hypothesis test using test purchases at 35 service stations across the state. He proposes to state his null hypothesis as follows: Ho: the mean price of gasoline is now greater than $3.00. Is this the correct null hypothesis?

6-4. In the gasoline example above, if the average selling price of gasoline has not changed, the transportation manager could find that the average price in the sample was (a) lower than $3.00, (b) higher than $3.00, or (c) either higher or lower than $3.00.

6-5. The transportation manager takes a sample of selling price from 36 service stations across the state. He finds that the average selling price in the sample was $3.10. It appears that the average (mean) price has gone up, but this could be just an accident of sampling. He wishes to know whether or not the price has indeed gone up. The sample had a standard deviation of five cents. A five percent lever of significance will be used. What is the null hypothesis? The alternate hypothesis? The critical value of the test statistic? The decision rule? Use the formula provided in the text and calculate the value of the test statistic. State your decision on the null hypothesis.

6-6. Professor Brown wonders if he is giving higher numerical scores on history exams than he has in the past. The university's statistics division tells him that the average score he has given, over the last thirty years, was a 75. His freshman class in the fall of 2009 had 36 students in it, and he decides to use this one class as a sample and conduct a hypothesis test at the one percent significance level. The average score in the class was 80 and the standard deviation of scores was 5 points. What is the null hypothesis? The alternate hypothesis? The critical value of the test statistic? The decision rule? Use the formula provided in the text and calculate the value of the test statistic. State your decision on the null hypothesis.

6-7. Your company pays a flat rate for expenses (meals and lodging) of $150 per day when your technical installers are on the road in the eastern U.S. This rate was based on actual expenses from a survey carried out five years ago. The vice president for sales and service insists that expenses have gone up and the rate should be increased. You agree to permit the VP conduct hypothesis test at the five percent significance level, using a sample of 50 actual expense records turned in just last month. The sample rate was $175 and the standard deviation of the sample was $15. What is the null hypothesis? The alternate hypothesis? The critical value of the test statistic? The decision rule? Use the formula provided in the text and calculate the value of the test statistic. State your decision on the null hypothesis.

6-8. Your company pays a flat rate for expenses (meals and lodging) of $150 per day when your technical installers are on the road in the eastern U.S. This rate was based on actual expenses from a survey carried out five years ago. The vice president for sales and service insists that expenses will be different in the new territory the company is expanding into – the Pacific Northwest. You agree to permit the VP conduct a hypothesis test at the five percent significance level, using a sample of 49 actual expense records in the new territory. The survey reveals that the rate in the sample of expenses in the Pacific Northwest was $200. The standard deviation of the sample was $25. What is the null hypothesis? The alternate hypothesis? The critical value of the test statistic? The decision rule? Use the formula provided in the text and calculate the value of the test statistic. State your decision on the null hypothesis.

Area under the Normal Curve from 0 to X

x	0.00	0.01	0.02	0.03	0.04	0.05	0.06	0.07	0.08	0.09
0.0	0.00000	0.00399	0.00798	0.01197	0.01595	0.01994	0.02392	0.02790	0.03188	0.03586
0.1	0.03983	0.04380	0.04776	0.05172	0.05567	0.05962	0.06356	0.06749	0.07142	0.07535
0.2	0.07926	0.08317	0.08706	0.09095	0.09483	0.09871	0.10257	0.10642	0.11026	0.11409
0.3	0.11791	0.12172	0.12552	0.12930	0.13307	0.13683	0.14058	0.14431	0.14803	0.15173
0.4	0.15542	0.15910	0.16276	0.16640	0.17003	0.17364	0.17724	0.18082	0.18439	0.18793
0.5	0.19146	0.19497	0.19847	0.20194	0.20540	0.20884	0.21226	0.21566	0.21904	0.22240
0.6	0.22575	0.22907	0.23237	0.23565	0.23891	0.24215	0.24537	0.24857	0.25175	0.25490
0.7	0.25804	0.26115	0.26424	0.26730	0.27035	0.27337	0.27637	0.27935	0.28230	0.28524
0.8	0.28814	0.29103	0.29389	0.29673	0.29955	0.30234	0.30511	0.30785	0.31057	0.31327
0.9	0.31594	0.31859	0.32121	0.32381	0.32639	0.32894	0.33147	0.33398	0.33646	0.33891
1.0	0.34134	0.34375	0.34614	0.34849	0.35083	0.35314	0.35543	0.35769	0.35993	0.36214
1.1	0.36433	0.36650	0.36864	0.37076	0.37286	0.37493	0.37698	0.37900	0.38100	0.38298
1.2	0.38493	0.38686	0.38877	0.39065	0.39251	0.39435	0.39617	0.39796	0.39973	0.40147
1.3	0.40320	0.40490	0.40658	0.40824	0.40988	0.41149	0.41308	0.41466	0.41621	0.41774
1.4	0.41924	0.42073	0.42220	0.42364	0.42507	0.42647	0.42785	0.42922	0.43056	0.43189
1.5	0.43319	0.43448	0.43574	0.43699	0.43822	0.43943	0.44062	0.44179	0.44295	0.44408
1.6	0.44520	0.44630	0.44738	0.44845	0.44950	0.45053	0.45154	0.45254	0.45352	0.45449
1.7	0.45543	0.45637	0.45728	0.45818	0.45907	0.45994	0.46080	0.46164	0.46246	0.46327
1.8	0.46407	0.46485	0.46562	0.46638	0.46712	0.46784	0.46856	0.46926	0.46995	0.47062
1.9	0.47128	0.47193	0.47257	0.47320	0.47381	0.47441	0.47500	0.47558	0.47615	0.47670
2.0	0.47725	0.47778	0.47831	0.47882	0.47932	0.47982	0.48030	0.48077	0.48124	0.48169
2.1	0.48214	0.48257	0.48300	0.48341	0.48382	0.48422	0.48461	0.48500	0.48537	0.48574
2.2	0.48610	0.48645	0.48679	0.48713	0.48745	0.48778	0.48809	0.48840	0.48870	0.48899
2.3	0.48928	0.48956	0.48983	0.49010	0.49036	0.49061	0.49086	0.49111	0.49134	0.49158
2.4	0.49180	0.49202	0.49224	0.49245	0.49266	0.49286	0.49305	0.49324	0.49343	0.49361
2.5	0.49379	0.49396	0.49413	0.49430	0.49446	0.49461	0.49477	0.49492	0.49506	0.49520
2.6	0.49534	0.49547	0.49560	0.49573	0.49585	0.49598	0.49609	0.49621	0.49632	0.49643
2.7	0.49653	0.49664	0.49674	0.49683	0.49693	0.49702	0.49711	0.49720	0.49728	0.49736
2.8	0.49744	0.49752	0.49760	0.49767	0.49774	0.49781	0.49788	0.49795	0.49801	0.49807
2.9	0.49813	0.49819	0.49825	0.49831	0.49836	0.49841	0.49846	0.49851	0.49856	0.49861
3.0	0.49865	0.49869	0.49874	0.49878	0.49882	0.49886	0.49889	0.49893	0.49896	0.49900
3.1	0.49903	0.49906	0.49910	0.49913	0.49916	0.49918	0.49921	0.49924	0.49926	0.49929
3.2	0.49931	0.49934	0.49936	0.49938	0.49940	0.49942	0.49944	0.49946	0.49948	0.49950
3.3	0.49952	0.49953	0.49955	0.49957	0.49958	0.49960	0.49961	0.49962	0.49964	0.49965
3.4	0.49966	0.49968	0.49969	0.49970	0.49971	0.49972	0.49973	0.49974	0.49975	0.49976
3.5	0.49977	0.49978	0.49978	0.49979	0.49980	0.49981	0.49981	0.49982	0.49983	0.49983
3.6	0.49984	0.49985	0.49985	0.49986	0.49986	0.49987	0.49987	0.49988	0.49988	0.49989
3.7	0.49989	0.49990	0.49990	0.49990	0.49991	0.49991	0.49992	0.49992	0.49992	0.49992
3.8	0.49993	0.49993	0.49993	0.49994	0.49994	0.49994	0.49994	0.49995	0.49995	0.49995
3.9	0.49995	0.49995	0.49996	0.49996	0.49996	0.49996	0.49996	0.49996	0.49997	0.49997
4.0	0.49997	0.49997	0.49997	0.49997	0.49997	0.49997	0.49998	0.49998	0.49998	0.49998

Source: Engineering Statistics Handbook (National Bureau of Standards)

http://www.itl.nist.gov/div898/handbook/eda/section3/eda3671.htm

Answers to chapter questions

6-1. In a census, every member of the population is measured. In a sample, only a part of the population is measured. This is done because measuring every variable in a population is likely to be too time-consuming, too expensive, or otherwise impractical or even impossible.

6-2. A mass of statistics, gathered over time, can come as close to proving something as humans are capable of. (For example: Legally drunk drivers are more likely to be involved in fatal accidents than sober drivers; states that pass concealed carry laws allowing honest people to carry concealed firearms have lower rates of violent crimes such as assault and rape than they did before the laws were passed, heavy smokers are more likely to have cardio-vascular disease than non-smokers.) But one sample, taken out of a large population, cannot prove anything. It can strongly suggest that something is or is not the case, but it cannot prove it. For example: Suppose this null hypothesis is tested. "The average weight of a sack of chicken feed from our plant is fifty pounds". If the test is run at the five percent significance level, out of one hundred times it is rejected, it will be rejected incorrectly five times, just because of a large accidental sampling error. If it is tested at the one percent level, it will be incorrectly rejected one out of a hundred times. This is called Type One Error.

6-3. This is not a testable null hypothesis. The null hypothesis must have an equal sign. Why? Because the distribution of sample means is centered on the population mean, whatever it is. It must be centered on a point. If the null hypothesis were stated as "the mean is greater than $3.00" the theoretical distribution of sample means used for the test could have an infinite number of possible locations, centered on prices ranging up from $3.00.

6-4. C. Due to sampling error the average price in the sample could be either higher or lower than $3.00.

6-5. Ho: The mean = $3
H1: The mean > $3
The critical value of the test statistic is 1.645
Reject Ho if Z > 1.645

$$Z = \frac{X - \mu}{S_{\bar{x}}}$$

$$Z = \frac{3.1 - 3}{.05/\sqrt{36}}$$

$$Z = \frac{.1}{.0083} = 12.04$$

The calculated value of Z is larger than 1.645, so the null hypothesis should be rejected.

6-6. Ho: The mean = 75
H1: The mean > 75
The critical value of the test statistic is 2.33
Reject Ho if Z > 2.33

$$Z = \frac{X - \mu}{S_{\bar{x}}}$$

$$Z = \frac{80-75}{5/\sqrt{36}}$$

$$Z = \frac{5}{5/6}$$

$$Z = 6.00$$

The calculated value of Z is larger than 2.33, the null hypothesis can be rejected. Either the students are studying more or Professor Brown is grading easier this year.

6-7. H_0: The mean = 150
H_1: The mean > 150
The critical value of the test statistic is 1.645
Reject H_0 if Z > 1.645

$$Z = \frac{\bar{X} - \mu}{S_{\bar{x}}}$$

$$Z = \frac{175 - 150}{15/\sqrt{50}}$$

$$Z = \frac{25}{15/7.071} = 11.79$$

The calculated value of Z is greater than 1.645, so the null hypothesis is rejected.

6-8. H_0: The mean = 150
H_1: The mean == 150 /
The critical value of the test statistic is 1.96
Reject H_0 if Z > 1.96 or < -1.96

$$Z = \frac{\bar{X} - \mu}{S_{\bar{x}}}$$

$$Z = \frac{200 - 150}{25/\sqrt{49}}$$

$$Z = \frac{50}{25/7} = 14.00$$

The calculated value of Z is greater than 1.96, so the null hypothesis is rejected.

Chapter 7

Hypothesis testing using two large samples

What this chapter will do for you.

This chapter will build on the material you learned in chapter eight about sampling theory and hypothesis testing. You will learn how to compare samples taken from two different populations to see if the means of the two populations are the same, or different.

A situation requiring a two-sample test

Let us suppose that your friend, the football coach from the previous chapter, was new to his job. He is straight out of college and has no data from previous years to draw on. Otherwise the situation is similar. The team he is taking over has traditionally worn white jerseys but he has read that black might be superior. As in the previous example, he will have a large number of scrimmages this spring. He decides to divide those scrimmages into 72 short scrimmage sessions of fifteen minutes each, and keep track, in each, of how many tackles with no gain are made by his defensive squad. At randomly selected times – mornings and evenings are mixed – the players will all wear white jerseys or black jerseys. Thirty six sessions his defensive players will wear white; in the other thirty six they will wear black.

If the players wore white jerseys from now on, for several years, the average number of tackles per 15 minutes of defensive play, with no gain, with white jerseys, would be represented by the symbol μ_w. That is the population mean. The coach does not know what it is and does not want to wait several years to find out.

If the players wore black jerseys from now on, for several years, the average number of tackles per 15 minutes of defensive play, with no gain, with black jerseys, would be represented by the symbol μ_b. That is the population mean. The coach does not know what it is and does not want to wait several years to find out.

What you will do to help him decide. You will use the 36 white-jersey scrimmage sessions as one sample, and the 36 black-jersey scrimmage sessions as a second sample.

The variation of sampling theory that will be applied to this problem.

In the previous chapter you learned that if every theoretically possible sample were taken from a population, the means of those samples, if plotted, would form a normal curve. That normal curve would be centered on the population mean. Something different but similar in concept occurs when a number of pairs of samples are taken out of a population.

Suppose, for example, that you have a population with a mean of 100. If you take a sample out of it, the mean of the sample, due to sampling accident, will almost never equal 100. It might be larger than 100. It might be smaller than 100.

If you flipped a coin it might come up heads or tails. If you flipped it again it might come up heads or tails. And so on. The odds never change from 50/50 for any particular flip of the coin. But, if you flipped the coin several hundred times, and recorded the result each time, you would find that close to half the time the coin would have come down heads, and half the time it would have come down tails.

In the same way, if you took a sample from a population, and calculated the mean of the sample, and then took a second sample, and calculated its mean, and subtracted the mean of the second sample from the mean of the first, you would get a difference. In theory, if you did this a very, very large number of times, you would find that half of the time the differences would be positive, and half of the time they would be negative. Most times the differences would be fairly small, whether positive of negative. A few times, the differences would be quite large, with the number of large positive differences equal to the number of large negative differences.

In fact – and we will just have to trust the mathematicians for this – these differences would be normally distributed, and this normal distribution of differences would be centered on zero. Half of the differences would be plus differences, and half would be negative. This situation is shown in the diagram below.

If two samples are taken from the same population, or if a sample is taken from one population, and a second sample is taken from an <u>identical</u> population, the result would be the same. The differences between samples would be half positive and half negative, and would be distributed normally. You can use the standard normal distribution to work with this distribution of differences, just as you did with the distribution of sample means in the previous chapter.

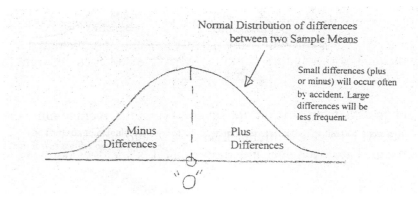

Let's assume now that the average number of no-gain stops per 15 minute defensive scrimmage segment was five when the squad wore black jerseys, and four when the wore white. It would certainly appear that black was the color of choice, but the coach wants a statistical test. Let's assume that the standard deviation within the first sample was two, the standard deviation within the second sample was three.

The null hypothesis, the alternate hypothesis, and the decision rule.

The null hypothesis would be that the mean of the black-jersey population would be the same as the mean of the white-jersey population.

H₀: $\mu_b = \mu_w$

The alternate hypothesis could be that the mean of the first sample mentioned (black-jersey) is greater than the mean of the second, that it is smaller than the mean of the second, or simply that the means are not equal. The three possibilities for the alternate hypothesis are shown below.

H₁: $\mu_b > \mu_w$ 　　　　H₁: $\mu_b < \mu_w$ 　　　　H₁: $\mu_b \neq \mu_w$

You choose the first, because that is what the coach is interested in and decide to use a five percent level of significance. Consequently, the decision rule would be "Reject Ho if Z > 1.645", the same value used in the example in the previous chapter. The figure below shows the region of rejection.

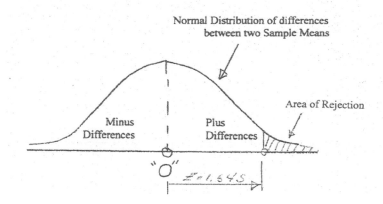

If you had decided to subtract the black-jersey mean of five from the white-jersey mean of four, the difference would have been negative, and the decision rule and the diagram would be reversed. It would not matter which you subtracted from which, as long as the calculations and the decision rule were consistent with each other – positive or negative. In this case we decide to subtract the white-jersey mean of four from the black-jersey mean of five and get a positive number.

Once again, the critical value of the test statistic is 1.645. The formula to calculate the value of the test statistic based on the facts of the problem is more complicated for the two sample problem.

$$Z = \frac{\overline{X}_1 - \overline{X}_2}{\sqrt{\dfrac{S_1^2}{n_1} + \dfrac{S_2^2}{n_2}}}$$

These variables are identified below. Numbers in parentheses are assumed for this problem, to illustrate its solution.

$\overline{X}_1$ is the mean of the first sample.   (5)

$\overline{X}_2$ is the mean of the second sample.   (4)

S_1^2 is the variance (the standard deviation squared) of the first sample. (2^2 or 4)

S_2^2 is the variance (the standard deviation squared) of the second sample. (3^2 or 9)

n_1 is the number of variables in the first sample.   (36)

n_2 is the number of variables in the second sample. (36)

$$Z = \frac{5 - 4}{\sqrt{\frac{4}{36} + \frac{9}{36}}}$$

$$Z = \frac{1}{\sqrt{.1111 + .25}}$$

$$Z = \frac{1}{\sqrt{.3611}}$$

$$Z = \frac{1}{.6009} = 1.664$$

This calculated value for the test statistics exceeds the critical value of the test statistic, and, according to the decision rule, would indicate that the null hypothesis should be rejected and the alternate hypothesis accepted. Your friend, the new coach, can be confident that jersey color makes a difference, and he needs to get new game jerseys in black ordered before the season begins.

A two tail test

A two tail test would be conducted in the same way as the two tail test in the previous chapter. The decision rule would establish a region of rejection on either side of the distribution of differences that is centered on zero. Different levels of significance could be established by referring to the Z chart. A ten percent two tail test would use the 1.645 value for example, while a five percent two tail would use the Z value from the chart for .4750, (Z = 1.96) which would put half of the five percent region of rejection in each tail.

What you have learned and what comes next.

You have learned how to extend your understanding of sampling theory to cover a situation where two samples are compared to judge whether or not they come from identical populations. In the next chapter you will learn to work with small samples.

Chapter Questions

7-1. If you were to take two samples out of the same population, the means of the two samples would be different, due to random sampling error. If you subtracted the mean of the second sample from the mean of the first sample, what percent of the time would the subtraction give you a positive number?

7-2. If you were to take two samples out of the same population, the means of the two samples would be different, due to random sampling error. If you subtracted the mean of the second sample from the mean of the first sample, what percent of the time would the subtraction give you a negative number?

7-3. If you were to take two samples out of two identical populations, the means of the two samples would be different, due to random sampling error. If you subtracted the mean of the second sample from the mean of the first sample, what percent of the time would the subtraction give you a positive number?

7-4. If you were to take two samples out of two identical populations, the means of the two samples would be different, due to random sampling error. If you subtracted the mean of the second sample from the mean of the first sample, what percent of the time would the subtraction give you a negative number?

7-5. The basketball coach at Northstar High School in Flint, Pennsylvania has a decision to make. His star center, Henry Hightower, who is now a senior, averaged 30 points per game over the last three years when the team played in its home gym, with a standard deviation of five points. He scored 20 points per game playing away from home, with a standard deviation of 10 points. In the course of these three seasons, the team played thirty home games and thirty away games. The coach has a new freshman recruit with an incredible high school scoring record. He is considering starting Henry Hightower in all home games, and starting the freshman sensation in away games, if a statistical hypothesis test at the ten percent level indicates that Henry does indeed score fewer points in away games. What is the null hypothesis? The alternate hypothesis? The critical value of the test statistic? The decision rule? Use the formula provided in the text and calculate the value of the test statistic. State your decision on the null hypothesis.

Area under the Normal Curve from 0 to X

X	0.00	0.01	0.02	0.03	0.04	0.05	0.06	0.07	0.08	0.09
0.0	0.00000	0.00399	0.00798	0.01197	0.01595	0.01994	0.02392	0.02790	0.03188	0.03586
0.1	0.03983	0.04380	0.04776	0.05172	0.05567	0.05962	0.06356	0.06749	0.07142	0.07535
0.2	0.07926	0.08317	0.08706	0.09095	0.09483	0.09871	0.10257	0.10642	0.11026	0.11409
0.3	0.11791	0.12172	0.12552	0.12930	0.13307	0.13683	0.14058	0.14431	0.14803	0.15173
0.4	0.15542	0.15910	0.16276	0.16640	0.17003	0.17364	0.17724	0.18082	0.18439	0.18793
0.5	0.19146	0.19497	0.19847	0.20194	0.20540	0.20884	0.21226	0.21566	0.21904	0.22240
0.6	0.22575	0.22907	0.23237	0.23565	0.23891	0.24215	0.24537	0.24857	0.25175	0.25490
0.7	0.25804	0.26115	0.26424	0.26730	0.27035	0.27337	0.27637	0.27935	0.28230	0.28524
0.8	0.28814	0.29103	0.29389	0.29673	0.29955	0.30234	0.30511	0.30785	0.31057	0.31327
0.9	0.31594	0.31859	0.32121	0.32381	0.32639	0.32894	0.33147	0.33398	0.33646	0.33891
1.0	0.34134	0.34375	0.34614	0.34849	0.35083	0.35314	0.35543	0.35769	0.35993	0.36214
1.1	0.36433	0.36650	0.36864	0.37076	0.37286	0.37493	0.37698	0.37900	0.38100	0.38298
1.2	0.38493	0.38686	0.38877	0.39065	0.39251	0.39435	0.39617	0.39796	0.39973	0.40147
1.3	0.40320	0.40490	0.40658	0.40824	0.40988	0.41149	0.41308	0.41466	0.41621	0.41774
1.4	0.41924	0.42073	0.42220	0.42364	0.42507	0.42647	0.42785	0.42922	0.43056	0.43189
1.5	0.43319	0.43448	0.43574	0.43699	0.43822	0.43943	0.44062	0.44179	0.44295	0.44408
1.6	0.44520	0.44630	0.44738	0.44845	0.44950	0.45053	0.45154	0.45254	0.45352	0.45449
1.7	0.45543	0.45637	0.45728	0.45818	0.45907	0.45994	0.46080	0.46164	0.46246	0.46327
1.8	0.46407	0.46485	0.46562	0.46638	0.46712	0.46784	0.46856	0.46926	0.46995	0.47062
1.9	0.47128	0.47193	0.47257	0.47320	0.47381	0.47441	0.47500	0.47558	0.47615	0.47670
2.0	0.47725	0.47778	0.47831	0.47882	0.47932	0.47982	0.48030	0.48077	0.48124	0.48169
2.1	0.48214	0.48257	0.48300	0.48341	0.48382	0.48422	0.48461	0.48500	0.48537	0.48574
2.2	0.48610	0.48645	0.48679	0.48713	0.48745	0.48778	0.48809	0.48840	0.48870	0.48899
2.3	0.48928	0.48956	0.48983	0.49010	0.49036	0.49061	0.49086	0.49111	0.49134	0.49158
2.4	0.49180	0.49202	0.49224	0.49245	0.49266	0.49286	0.49305	0.49324	0.49343	0.49361
2.5	0.49379	0.49396	0.49413	0.49430	0.49446	0.49461	0.49477	0.49492	0.49506	0.49520
2.6	0.49534	0.49547	0.49560	0.49573	0.49585	0.49598	0.49609	0.49621	0.49632	0.49643
2.7	0.49653	0.49664	0.49674	0.49683	0.49693	0.49702	0.49711	0.49720	0.49728	0.49736
2.8	0.49744	0.49752	0.49760	0.49767	0.49774	0.49781	0.49788	0.49795	0.49801	0.49807
2.9	0.49813	0.49819	0.49825	0.49831	0.49836	0.49841	0.49846	0.49851	0.49856	0.49861
3.0	0.49865	0.49869	0.49874	0.49878	0.49882	0.49886	0.49889	0.49893	0.49896	0.49900
3.1	0.49903	0.49906	0.49910	0.49913	0.49916	0.49918	0.49921	0.49924	0.49926	0.49929
3.2	0.49931	0.49934	0.49936	0.49938	0.49940	0.49942	0.49944	0.49946	0.49948	0.49950
3.3	0.49952	0.49953	0.49955	0.49957	0.49958	0.49960	0.49961	0.49962	0.49964	0.49965
3.4	0.49966	0.49968	0.49969	0.49970	0.49971	0.49972	0.49973	0.49974	0.49975	0.49976
3.5	0.49977	0.49978	0.49978	0.49979	0.49980	0.49981	0.49981	0.49982	0.49983	0.49983
3.6	0.49984	0.49985	0.49985	0.49986	0.49986	0.49987	0.49987	0.49988	0.49988	0.49989
3.7	0.49989	0.49990	0.49990	0.49990	0.49991	0.49991	0.49992	0.49992	0.49992	0.49992
3.8	0.49993	0.49993	0.49993	0.49994	0.49994	0.49994	0.49994	0.49995	0.49995	0.49995
3.9	0.49995	0.49995	0.49996	0.49996	0.49996	0.49996	0.49996	0.49996	0.49997	0.49997
4.0	0.49997	0.49997	0.49997	0.49997	0.49997	0.49997	0.49998	0.49998	0.49998	0.49998

Source: Engineering Statistics Handbook (National Bureau of Standards)

http://www.itl.nist.gov/div898/handbook/eda/section3/eda3671.htm

Answers to chapter questions

7-1. Fifty percent

7-2. Fifty percent

7-3. Fifty percent

7-4. Fifty percent.

7-5. Ho: $\mu_1 = \mu_2$

H1: $\mu_1 > \mu_2$
The critical value of the test statistic is 1.28
Reject Ho if Z > 1.28

$$Z = \frac{\overline{X}_1 - \overline{X}_2}{\sqrt{\frac{S_1^2}{n_1} + \frac{S_2^2}{n_2}}}$$

$$Z = \frac{30 - 20}{\sqrt{\frac{5^2}{30} + \frac{10^2}{30}}}$$

$$Z = \frac{10}{\sqrt{.8333 + 3.3333}}$$

$$Z = \frac{10}{\sqrt{4.16667}}$$

$$Z = \frac{10}{2.0412} = 4.899$$

The calculated value of the test statistic exceeds 1.28. The null hypothesis would be rejected.

Chapter 8

Hypothesis tests using small samples

What this chapter will do for you.

Chapters 6 and 7 dealt with "large" samples, that is, samples where the number of variables measured in the sample was equal to or larger than thirty. The hypothesis testing theory underlying small samples is similar, but the critical values of the test statistic (called the "t" statistic) will vary depending on the size of the sample or samples taken. This chapter will build on the material you learned in previous chapters about sampling theory and hypothesis testing. You will learn, first, how use a small sample to determine if you should believe that a population mean has changed (similar to the situation in Chapter 6), second, to compare samples taken from two different populations to see if the means of the two populations are the same (similar to the situation in Chapter 7), and, third, to apply a test unique to small samples, where descriptive sample numbers are available in matched pairs.

Hypothesis tests using small samples.

Generally, if a sample size of thirty or more is used, and ratio data is being analyzed, the Z statistic will be used for hypothesis testing. This has been demonstrated in Chapters 8 and 9. For smaller samples however a different test statistic should be used: the "t" statistic. Three different t tests may be used in small sample hypothesis testing, depending on the problem and the data available.

The forms that the null hypothesis would take under these three kinds of problems

A one sample test can be conducted to determine if a null hypothesis stating that the population mean is equal to some specified value, should be accepted or rejected. This would be quite similar to the large sample test in Chapter 6. The null hypothesis would be stated as follows:

H_0: μ = some particular value

A two sample test can be conducted to determine if the means of two populations are equal. This would be quite similar to the two sample Z test conducted in Chapter 7, although the formula to calculate the value of the test statistic would be different. The null hypothesis would look like this:

H_0: $\mu_1 = \mu_2$

A comparison of paired observations (dependent samples) can be carried out to determine if a population has changed, or if two populations have different means. This procedure is different from anything considered so far. In this procedure, an observation is taken from one population, and a matched observation (the other part of a pair) is taken from another population, and this is done for a number of pairs, to measure the average difference between the two items in the pairs.

Is this confusing? Let's illustrate by example. Suppose you are a safety director for a railroad, and you have just supervised an improvement of the warning signals at a large number of road crossings. You wish to determine if accidents have been reduced. You could simply compare samples taken before and after the upgrade, of the total number of accidents, as outlined in the second type of problem described above. But mathematicians tell us that a more accurate judgment can be made if we compare the number of accidents, before and after, at particular, individual crossings. For a particular crossing, the before and after numbers would be one pair. Numbers of accidents could be found for a small group of crossings, and the before and after numbers obtained for each crossing. Then, a test of paired differences, sometimes called a test of dependent samples, could be carried out.

In this case the null hypothesis would state that the average difference between data points in all pairs (all crossings) - the "mean of the differences" - before and after the improvement program, would be zero. The null hypothesis would be stated as follows:

H_0: $\mu_d = 0$

We will now discuss each of these three types of problems in order, starting with the one sample problem. Because of the length of this chapter, the t table that will be used is on the next page, as well as at the end of the chapter.

The formula for calculating the value of the test statistic is almost identical to the formula you have already had practice in using, in a one sample, large sample test in Chapter 6. It would be stated as follows:

$$t = \frac{\overline{X} - \mu}{S_{\overline{X}}}$$

It is the same as the formula you used before, except that t is substituted for Z.

As before, the alternate hypothesis could be that the mean is now greater than some established (or former) value, smaller than that value, or simply not equal to that value (it could be either larger or smaller).

As is our practice, let us illustrate in the clearest way possible, by example, following the Table of t values on the next page..

Values of Student's t distribution with degrees of freedom in column headings
Probability of exceeding the critical value

	0.10	0.05	0.025	0.01	0.005	0.001
1.	3.078	6.314	12.706	31.821	63.657	318.313
2.	1.886	2.920	4.303	6.965	9.925	22.327
3.	1.638	2.353	3.182	4.541	5.841	10.215
4.	1.533	2.132	2.776	3.747	4.604	7.173
5.	1.476	2.015	2.571	3.365	4.032	5.893
6.	1.440	1.943	2.447	3.143	3.707	5.208
7.	1.415	1.895	2.365	2.998	3.499	4.782
8.	1.397	1.860	2.306	2.896	3.355	4.499
9.	1.383	1.833	2.262	2.821	3.250	4.296
10.	1.372	1.812	2.228	2.764	3.169	4.143
11.	1.363	1.796	2.201	2.718	3.106	4.024
12.	1.356	1.782	2.179	2.681	3.055	3.929
13.	1.350	1.771	2.160	2.650	3.012	3.852
14.	1.345	1.761	2.145	2.624	2.977	3.787
15.	1.341	1.753	2.131	2.602	2.947	3.733
16.	1.337	1.746	2.120	2.583	2.921	3.686
17.	1.333	1.740	2.110	2.567	2.898	3.646
18.	1.330	1.734	2.101	2.552	2.878	3.610
19.	1.328	1.729	2.093	2.539	2.861	3.579
20.	1.325	1.725	2.086	2.528	2.845	3.552
21.	1.323	1.721	2.080	2.518	2.831	3.527
22.	1.321	1.717	2.074	2.508	2.819	3.505
23.	1.319	1.714	2.069	2.500	2.807	3.485
24.	1.318	1.711	2.064	2.492	2.797	3.467
25.	1.316	1.708	2.060	2.485	2.787	3.450
26.	1.315	1.706	2.056	2.479	2.779	3.435
27.	1.314	1.703	2.052	2.473	2.771	3.421
28.	1.313	1.701	2.048	2.467	2.763	3.408
29.	1.311	1.699	2.045	2.462	2.756	3.396
30.	1.310	1.697	2.042	2.457	2.750	3.385
35.	1.306	1.690	2.030	2.438	2.724	3.340
40.	1.303	1.684	2.021	2.423	2.704	3.307
45.	1.301	1.679	2.014	2.412	2.690	3.281
50.	1.299	1.676	2.009	2.403	2.678	3.261

Source: Engineering Statistics Handbook (National Bureau of Standards)

http://www.itl.nist.gov/div898/handbook/eda/section3/eda3672.htm

A hypothesis test using one small sample.

You have a friend, Joe Jones, who has, for years, used organic farming techniques (no pesticides or herbicides, no chemical fertilizers) on the fields where he grows beans. For years, the yield from his fields has been 30 bushels of beans per acre. This year, he began a practice of adding small amounts of fertilizer from his chicken flock to the fields. He wonders if he should believe that this practice has increased his bean yields. Large samples are not practical, but he does manage to take 28 samples from different parts of his fields, and calculates that the average yield in the sample this year is 33 bushels per acre. It does appear that yields have increased, but is this apparent increase just an accident of sampling, or does it point to a real, overall increase in bean yields?

The standard deviation of his sample was 2.1 bushels. He wishes to use a five percent hypothesis test.

The null hypothesis would be stated as follows:

Ho: $\mu = 40$

The alternate hypothesis would be stated as follows:

H_1: $\mu > 40$

Since Joe wants to know if he should believe that the yield is larger, you will be working with a one tail test. The region of rejection will be out in the right tail of the distribution of possible sample means. (If he simply wanted to know if the yield was now different - not equal to 30 bushels per acre - you would have a two tail test.)

Just as you learned to look up the critical value of the Z statistic in the Z table, depending on the level of significance, you will need to look up the critical value of the t statistic in a t table, so you can set down the decision rule. The t table is differently organized from the Z table however. Why is this?

If we took a census of a population – that is, if we measured every item in the population – we could calculate its mean exactly. When we take a sample however, and try to guess from it what the population mean is, we become less certain. If the sample mean becomes small (less than 30) the uncertainty becomes greater. To illustrate, what if we took a sample of one? We would be trying to guess just from one sample value what the population mean was. What if we took a sample of two? We could average the two sample values and guess that average could be the population mean, but it is not likely we would be correct. As we go down in sample size from thirty then, the uncertainty increases.

Mathematicians use a term "degrees of freedom" as an indication of the degree of uncertainty. For our purposes here we need to remember only that the number in the sample (n), minus one, equals the degrees of freedom in the problem. You could think of it this way: If one sample value is the anchor of a group of numbers, all the others are free to vary in value with respect to each other and with respect to the anchor variable. The number of sample variables free to vary would be equal to the degrees of freedom.

When you examine the t table, you note that the possible critical values of the test statistic, t, are arranged in columns. If you were using a one tail test at the five percent significance level, you would look at the row of headings labeled "One Tail", and find the needed value for your decision rule somewhere in the column under the .05% column heading. But where under that column heading?

On the left side of the t table you find a column headed "df". This stands for degrees of freedom. In our problem the number in the sample was 28. If we subtract one from 28 we have 27 degrees of freedom. Going down the "five percent" column, to 27 degrees of freedom, we see that the critical value of t for our problem is 1.703.

Values of Student's t distribution with degrees of freedom in column headings

Probability of exceeding the critical value

	0.10	0.05	0.025	0.01	0.005	0.001
1.	3.078	6.314	12.706	31.821	63.657	318.313
2.	1.886	2.920	4.303	6.965	9.925	22.327
3.	1.638	2.353	3.182	4.541	5.841	10.215
4.	1.533	2.132	2.776	3.747	4.604	7.173
5.	1.476	2.015	2.571	3.365	4.032	5.893
6.	1.440	1.943	2.447	3.143	3.707	5.208
7.	1.415	1.895	2.365	2.998	3.499	4.782
8.	1.397	1.860	2.306	2.896	3.355	4.499
9.	1.383	1.833	2.262	2.821	3.250	4.296
10.	1.372	1.812	2.228	2.764	3.169	4.143
11.	1.363	1.796	2.201	2.718	3.106	4.024
12.	1.356	1.782	2.179	2.681	3.055	3.929
13.	1.350	1.771	2.160	2.650	3.012	3.852
14.	1.345	1.761	2.145	2.624	2.977	3.787
15.	1.341	1.753	2.131	2.602	2.947	3.733
16.	1.337	1.746	2.120	2.583	2.921	3.686
17.	1.333	1.740	2.110	2.567	2.898	3.646
18.	1.330	1.734	2.101	2.552	2.878	3.610
19.	1.328	1.729	2.093	2.539	2.861	3.579
20.	1.325	1.725	2.086	2.528	2.845	3.552
21.	1.323	1.721	2.080	2.518	2.831	3.527
22.	1.321	1.717	2.074	2.508	2.819	3.505
23.	1.319	1.714	2.069	2.500	2.807	3.485
24.	1.318	1.711	2.064	2.492	2.797	3.467
25.	1.316	1.708	2.060	2.485	2.787	3.450
26.	1.315	1.706	2.056	2.479	2.779	3.435
27.	1.314	1.703	2.052	2.473	2.771	3.421

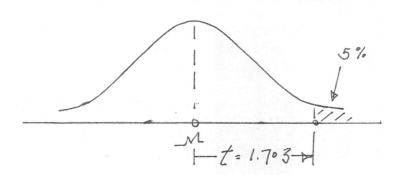

The decision rule then becomes: "Reject H_0 if t > 1.703"

The next step is to use the information we have from our sample to calculate a value of t (the test statistic) and compare it with this critical value of the test statistic obtained from the t table.

$$t = \frac{\bar{X} - \mu}{S_{\bar{x}}}$$

$$t = \frac{33 - 30}{2.1 / \sqrt{28}}$$

$$t = \frac{3}{2.1 / 5.2915}$$

$$t = \frac{3}{.39686} = 7.55934$$

It is customary to round off the calculated value of the test statistic to three places:

t = 7.559

What is the decision on the null hypothesis? Since 7.559 is not larger than 1.703 the decision rule tells us that we should not reject the null hypothesis that the mean yield is still 30 bushels per acre, and accept the alternate hypothesis that it is now larger.

A hypothesis test using two small samples.

The theory of using two small samples to determine if we should believe that two population means are the same is the same theory that underlies use of two large samples. The primary difference from the two large sample test (aside from using the t table for the critical value of the test statistic) is a more complicated formula for calculating the value of the test statistic to compare with the critical value from the table.

The value of the t statistic is calculated using the formula below.

$$t = \frac{\bar{X}_1 - \bar{X}_2}{\sqrt{S_p^2 (1/n_1 + 1/n_2)}}$$

n_1 is the number of the first sample.
n_2 is the number of the second sample.
$\bar{X}_1$ is the mean of the first sample.
$\bar{X}_2$ is the men of the second sample.

In these tests the assumption is made that the variances of the two populations are equal. If this is not the case a more complicated formula, found in most texts, can be used, however the procedure below relies on the common and convenient assumption that the two populations have the same variance.

S_p^2 is the "pooled variance", and is calculated using the formula below.

$$S_p^2 = \frac{(n_1 - 1)(S_1^2) + (n_2 - 1)(S_2^2)}{n_1 + n_2 - 2}$$

S_1^2 is the variance of the first sample (the standard deviation squared)
S_2^2 is the variance of the second sample (the standard deviation squared

In a two sample problem, just as with large samples, the null hypothesis is that two population means are equal. The alternate hypothesis could be that the mean of the first sample mentioned is greater than the mean of the second, that it is smaller than the mean of the second, or simply that the means are not equal. The three possibilities for the alternate hypothesis are shown below.

$H_1: \mu_1 > \mu_2$ $\qquad$ $H_1: \mu_1 < \mu_2$ $\qquad$ $H_1: \mu_1 \neq \mu_2$

A two sample problem.

Suppose an auto insurance executive has been told that there are more accidents in Oak Grove than in Maple City. The two cities have the same population and the same number of cars, and she finds this difficult to believe. She decides on a hypothesis test to determine if the company should indeed operate on the basis that, over a long period of time, there will be more accidents in city Oak Grove.

It would be impractical to find the population averages. That could take years, depending on a judgment as to how many years it would take to constitute a population of accidents. A decision is made to count the number of accidents per day for a small number of days. This results in the sampling data shown in the table below.

	Maple City	Oak Grove
Average number of accidents per day	40	43
Sample standard deviation	3	5
Sample size	21	26

The executive wishes to determine, at the one percent level of confidence, if Oak Grove actually has more accidents per day.

The first question might be, "What is the critical value of the test statistic?" To find this value we would use a standard t table. The number of degrees of freedom is sample size one plus sample size two, minus 2. We first locate the column heading of .01 in the one tail row of headings, then we read down to a "df" of 45. (This is 21 + 26 – 2.) The critical value of t is 2.412.

Values of Student's t distribution with degrees of freedom in column headings
Probability of exceeding the critical value

df	0.10	0.05	0.025	0.01	0.005	0.001
45	1.301	1.679	2.014	2.412	2.690	3.281

The null hypothesis would be stated as follows:

H_0: $\mu_{MC} = \mu_{OG}$

The alternate hypothesis would be stated as follows:

H_1: $\mu_{MC} < \mu_{OG}$

The decision rule would be "Reject Ho if t < - 2.412

Why a minus sign, and why "less than"? If we subtract the second sample mean from the first, we will get a negative number. And that is logical, because, if there are more accidents in Oak Grove, that would be expected. Here, the region of rejection is in the left tail of the normal curve which shows the normal distribution of accidental differences that would exist between sample means if the population means were actually equal.

The first step in calculating the value of the test statistic is to calculate the pooled variance.

$$S_p^2 = \frac{(n_1 - 1)(S_1^2) + (n_2 - 1)(S_2^2)}{n_1 + n_2 - 2}$$

$$S_p^2 = \frac{(21 - 1)(3^2) + (26 - 1)(5^2)}{45}$$

$$S_p^2 = \frac{(21 - 1)(9) + (26 - 1)(25)}{45}$$

$$S_p^2 = \frac{180 + 625}{45} = 17.8889$$

Now that we have the pooled variance we can plug it into the formula to calculate the value of the test statistic.

$$t = \frac{\overline{X_1} - \overline{X_2}}{\sqrt{S_p^2(1/n_1 + 1/n_2)}}$$

$$t = \frac{40 - 43}{\sqrt{17.8889(1/21 + 1/26)}}$$

$$t = \frac{-3}{\sqrt{17.8889(.0861)}}$$

$$t = \frac{-3}{\sqrt{1.5402}}$$

$$t = \frac{-3}{1.2410} = -2.417$$

Since negative 2.417 is less than minus 2.412 the null hypothesis is rejected.

A hypothesis test using paired observations.

There are two formulas that are used in this kind of problem. The formula below is used to calculate the value of the test statistic. Before we can use it, we must calculate the standard deviation of the differences, Sd.

$$t = \frac{\overline{d}}{S_d/\sqrt{n}}$$

Sd, the standard deviation of the differences between the two items of the pairs, is calculated using the formula below.

$$S_d = \sqrt{\frac{\Sigma d^2 - (\Sigma d)^2/n}{n-1}}$$

The null hypothesis will be that the mean difference is zero (that there is no real difference in the size of the variables as a result of some intervention.)

The alternate hypothesis could be that the mean difference is negative, it is positive, or it is simply not equal to zero.

$H_1: \mu_d > 0$ $\qquad\qquad$ $H_1: \mu_d < 0$ $\qquad\qquad$ $H_1: \mu_d \neq 0$

A dependent sample (paired differences) problem.

Suppose an inventor comes to an agricultural experiment station claiming to have developed a substance that will cause tomato vines to bear larger numbers of tomatoes in a growing season than the fertilizer presently used. The agricultural scientists agree to test the old fertilizer and the new substance using nine kinds of tomato vine. A pair of Big Boy tomato vines is obtained and planted. The old fertilizer is put on one of the vines and the new substance on the other. Next a pair of Better Girl tomato vines is obtained and planted, and the old fertilizer is put on one and the new substance on the other. This is done for an additional seven kinds of tomato vine, for a total of nine pairs in all. The scientists wish to determine if there is a <u>difference</u> between the old fertilizer and the new substance. This means we have a two tail test, since a difference could occur either way.

The null hypothesis would state that the mean of the differences (considering all possible pairs of tomato plants) is zero.

$H_0: \mu_d = 0$

The alternate hypothesis would be stated as follows:

$H_1: \mu_d \neq 0$

We need to determine the critical value of the test statistic. The degrees of freedom would be the number of pairs minus one, or eight. Let us assume that we want to use a five percent level of significance. When we refer to the t table, using the .05 two tail column, and eight degrees of freedom (one less than the number of pairs), and look for one half of the region of rejection (one half of five percent or .025) in each tail, we find that the critical value of the test statistic is 2.306.

The decision rule would read "Reject H_0 if t is greater than 2.306 or less than a minus 2.306.

"Reject H_0 if t > 2.306 or < - 2.306"

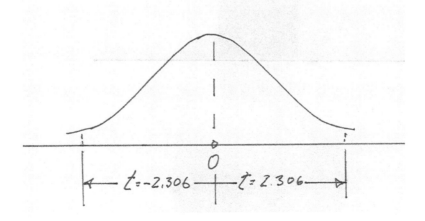

To work this problem we are given the following sample data.

	Pair								
	1	2	3	4	5	6	7	8	9
Old Fertilizer	8	10	15	13	15	25	17	11	20
New Substance	10	8	20	10	15	20	18	17	18
Differences	+2	-2	+5	-3	0	-5	+1	+6	-2
The differences Squared	4	4	25	9	0	25	1	36	4

The sum of the differences will equal +2.
The sum of the squares of the differences will equal 108
The average difference will be 2 / 9 or .22

First we must calculate the standard deviation using the formula listed earlier.

$$S_d = \sqrt{\frac{\Sigma d^2 - (\Sigma d)^2 / n}{n - 1}}$$

$$S_d = \sqrt{\frac{108 - (2)^2 / 9}{9 - 1}}$$

$$S_d = \sqrt{\frac{108 - 4/9}{9 - 1}}$$

$$S_d = \sqrt{\frac{108 - .4444}{8}}$$

$$S_d = \sqrt{13.44445} = 3.6667$$

This value for the standard deviation of the differences can now be used to calculate the value of the t statistic.

$$t = \frac{\bar{d}}{S_d / \sqrt{n}}$$

$$t = \frac{.22}{3.6667 / \sqrt{9}}$$

$$t = \frac{.22}{1.2222} = 0.180$$

Since 0.180 is not greater than the critical value of the test statistic (2.306) and is not less than a minus 2.306, the null hypothesis is not rejected.

What you have learned and what comes next

Many real world problems must be worked with only small samples. You have learned to test for a population mean, to see if you should believe that the means of two populations are the same or different, and to conduct tests using dependent samples (paired differences). The topic of the next chapter is Analysis of Variance, a unique procedure that permits comparison of three or more samples.

Problems for this chapter.

Using one sample to make conclusions about a population mean.

8– one sample-1. The Sunny Meadow Dairy Farm has traditionally sold its product exclusively under its own brand name. Its monthly average using this method of marketing exclusively has been $89,000. Beginning in January it has continued sales under its own brand name, as before, and, using the marginal costing concept, has begun to supply the Grizzly Bear Supermarket Chain with products the chain sells under its house brand. Profits for the first nine months of this year have averaged $97,000. The standard deviation of this nine month sample was calculated to be $10,000. At the five percent level of significance should the farm manager conclude that monthly profits are increased using this new marketing concept? What is the null hypothesis? What is the alternate hypothesis? What is the critical value of the test statistic? What is the decision rule? Calculate the value of the test statistic. State whether the null hypothesis should be rejected.

8- one sample- 2. The Affordable Construction Company, a homebuilder, competes primarily on the basis of price, rather than amenities. Floor plans are standardized, variations are limited. The chief purchasing officer shops extensively for lumber and other materials. Sometimes the lumber is not top grade, or even close, and the production foreman figures this not only increases labor hours but also results in more scrap than if better materials were used. The historical labor and materials cost for their basic three bedroom, two bath house, adjusted to today's dollars, is $75,000.

The CEO has wondered whether or not a switch to higher quality materials, although it would increase materials prices, might decrease overall labor and materials costs. This year, the purchasing officer was instructed to continue being price conscious but to seek out quality supplies and to establish buying relationships, if possible, with suppliers who could be depended on to deliver high quality materials to each job site. The company has completed seven basic houses under the new purchasing system. The average cost was $70,000 and the standard deviation of this sample was $5,000. At the five percent significance level can the CEO conclude that the new purchasing policy has reduced overall labor and materials cost? What is the null hypothesis? What is the alternate hypothesis? What is the critical value of the test statistic? What is the decision rule? Calculate the value of the test statistic. State whether the null hypothesis should be rejected.

8-one sample-3. Fred's Flower Shop sells flowers for birthdays and other special occasions, but depends heavily for yearly sales on just three holidays: Valentines Day, Mothers Day, and Christmas. Sales on each of these holidays have averaged $100,000 over the years. Fred's has now begun to advertise on the small city's talk radio station during the week prior to each of these holidays. This year, sales have averaged $120,000. Since the local economy is basically unchanged this year, management thinks that advertising has increased sales. The standard deviation of this small sample is $25,000. At the one percent significance level, should management conclude that this is the case? What is the null hypothesis? What is the alternate hypothesis? What is the critical value of the test statistic? What is the decision rule? Calculate the value of the test statistic. State whether the null hypothesis should be rejected.

8 -one sample- 4. Indiana Industries assembles a state of the art personal computer, using mail order parts, in 97 minutes. A new method has been introduced. This week, following practice runs to get accustomed to the new method, Hotwire technicians assembled twelve of these computers using the new method. The average time was 85 minutes. The standard deviation of this sample was 13 minutes. Should Hotwire management conclude, at the one percent significance level, that assembly time has been reduced? What is the null hypothesis? What is the alternate hypothesis? What is the critical value

of the test statistic? What is the decision rule? Calculate the value of the test statistic. State whether the null hypothesis should be rejected.

8-one sample- 5. The Apollo Fitness Center has used traditional workout programs for its clients. Average weight loss for the client's first year was 27 pounds. Slightly more than a year ago, the center added, for each client, dietary and nutritional advice. For the first seven clients completing the new program the average weight loss was 45 pounds. The standard deviation of this sample was 17 pound. At the one percent significance level has the new program caused the typical client to shed more pounds. What is the null hypothesis? What is the alternate hypothesis? What is the critical value of the test statistic? What is the decision rule? Calculate the value of the test statistic. State whether the null hypothesis should be rejected.

8- one sample- 6. Consider the information in problem five above. At the one percent significance lever has the average amount of weight loss changed? What is the null hypothesis? What is the alternate hypothesis? What is the critical value of the test statistic? What is the decision rule? Calculate the value of the test statistic. State whether the null hypothesis should be rejected.

8- one sample- 7. Enoch Jones raises vegetables to sell in a farmers' market. Tomatoes are an important item in season (typically July through September). He has kept careful records and knows he has had an average yield of 5 pounds per plant over the season, using commercial chemical fertilizer. His neighbor keeps large numbers of chickens and has a problem disposing of the residue from the chickens, as well as the necessity of paying a local college student to scoop out the chicken houses weekly. The neighbor has offered the fertilizer to Enoch free, if he will clean out the chicken houses on a regular basis. Before agreeing to do this entirely, Enoch has conducted a test this year. Using the chicken fertilizer on just fifteen plants over the season, he has found that they averaged 6.5 pounds of tomatoes. This sample had a standard deviation of .5 pounds. At the five percent significance level, should Enoch conclude that yields have increased? What is the null hypothesis? What is the alternate hypothesis? What is the critical value of the test statistic? What is the decision rule? Calculate the value of the test statistic. State whether the null hypothesis should be rejected.

Using two small samples to determine if the means of two populations are the same.

8-two sample-1. Central City's Oil Change Depot has two bays, identically equipped. The store has been open for 18 days. Jim Jones, in bay one has averaged 21 oil changes per day. For this sample, the standard deviation was 3. Sammy Samson, running bay two has averaged 23 oil changes per day, with a standard deviation of 4. The manager wishes to use a small sample hypothesis test, at the five percent level, to see if there is a difference in the output of the two employees. What is the null hypothesis? What is the alternate hypothesis? What is the critical value of the test statistic? What is the decision rule? Calculate the value of the test statistic. State whether or not the null hypothesis should be rejected.

8- two sample-2. Professor Sandstone's statistics class has seven female students and nine male students. On the final examination the average grade of the female students was 79 percent with a standard deviation of 15 percent. The average score of the male students was 83, with a standard deviation of 25 percent. At the five percent significance level should the professor conclude that there is a difference in male and female scores overall? What is the null hypothesis? What is the alternate hypothesis? What is the critical value of the test statistic? What is the decision rule? Calculate the value of the test statistic. State whether the null hypothesis should be rejected.

8- two sample- 3. Last year, on the three primary flower selling holidays, a florist shop averaged sales of $100,000, with a standard deviation of $1,000. This year, the average was $120,000 with, oddly enough, a standard deviation of $1,000. At the one percent significance lever, should management conclude that sales have increased? What is the null hypothesis? What is the alternate hypothesis? What is the critical value of the test statistic? What is the decision rule? Calculate the value of the test statistic. State whether the null hypothesis should be rejected.

8-two sample-4. Consider the Hotwire Industries problem above. Suppose the 97 minute time was based on a sample of 29 computers, and the standard deviation of this sample was ten minutes. Should Hotwire management conclude, at the one percent significance level, that assembly time has been reduced? What is the null hypothesis? What is the alternate hypothesis? What is the critical value of the test statistic? What is the decision rule? Calculate the value of the test statistic. State whether the null hypothesis should be rejected.

Problems using dependent samples (paired differences).

8-paired differences -1. The village of Maple Grove has four intersections, which already have traffic lights, where there have been a large number of accidents resulting when cars turn left in front of oncoming cars (both of which have green lights). The village council decides to replace the traffic lights at these intersections with new signals. These signals have an extra light box, labeled "Left Turn". At all times, except when cars have a green left turn arrow showing in this box, there is a blinking red light. The council wished to determine, at the one percent significance level, if it should conclude that the new signals have reduced the number of accidents. What is the null hypothesis? What is the alternate hypothesis? What is the critical value of the test statistic? What is the decision rule? Calculate the value of the test statistic. State whether the null hypothesis should be rejected. The before and after data are shown below.

Intersection	1	2	3	4
Number before change	2	6	7	3
Number after change	3	3	2	1

8 -paired differences- 2. Stan Simpson, coach of a second division NFL team believes that the team can win more games if his safeties and corners will become "ball hawks". In the off season, he put six of his players (four starters and two subs) through a film training exercise, where they were required to sit and watch, for seven hours a day over a period of 12 weeks, films of opposing quarterbacks making their reads and releasing the ball. He wished to evaluate this exercise by comparing, at the one percent significance level, their performance over the course of the season. What is the null hypothesis? What is the alternate hypothesis? What is the critical value of the test statistic? What is the decision rule? Calculate the value of the test statistic. State whether the null hypothesis should be rejected.

Results are shown below.

Player	Able	Baker	Charlie	Dobson	Emmerson	Finegold
Interceptions Last year	2	3	5	2	1	1
Interceptions This year	3	5	4	2	3	2

8- paired differences- 3. Consider the Jones Tomato Farm in one-sample problem seven above. Suppose Mr. Jones plants five different varieties of tomato vines (early bearers, mid-season bearers, and late bearers and tries old and new fertilizers on just 10 vines, two of each of the five types he typically plants. The results are given below. He wishes to determine at the five percent significance level if the new fertilizer increases yields. What is the null hypothesis? What is the alternate hypothesis? What is the critical value of the test statistic? What is the decision rule? Calculate the value of the test statistic. State whether the null hypothesis should be rejected.

Vine	Redbird Early	Mammoth Mid-season	Big Beef	Linger Longer
Yield (old)	4.2	5.0	4.9	3.9
Yield (new)	4.3	5.3	5.4	3.7

8- paired differences- 4. The police chief has been frustrated by the number of violent assaults against women that has plagued the community of Central City (the home of a University of 20,000 enrollment) during the academic year. A local chapter of the National Rifle Association has come to him offering to make available to the people of the community a self defense course consisting of (1) situational awareness, (2) basic precautions (regarding being on the street alone at odd hours, etc.), and (3) fundamentals of self defense with a firearm as made possible by the state's recent passage of a concealed carry permit law. This class was made available to all who wished to attend in August and September 2008. The police chief wishes to report the results to the village council. He asks your help in conducting a hypothesis test at the one percent significance level to determine if the number of violent assaults has been reduced. What is the null hypothesis? What is the alternate hypothesis? What is the critical value of the test statistic?

What is the decision rule? Calculate the value of the test statistic. State whether the null hypothesis should be rejected.

Month	Sept	Oct	Nov	Dec	Jan	Feb	Mar	Apr	May
Assaults (2007)	5	4	3	3	2	5	5	4	2
Assaults (2008)	2	3	2	4	4	2	3	1	0

$\Sigma d = -12$ $\Sigma d^2 = 42$

Values of Student's t distribution with degrees of freedom in column headings
Probability of exceeding the critical value

df	0.10	0.05	0.025	0.01	0.005	0.001
1.	3.078	6.314	12.706	31.821	63.657	318.313
2.	1.886	2.920	4.303	6.965	9.925	22.327
3.	1.638	2.353	3.182	4.541	5.841	10.215
4.	1.533	2.132	2.776	3.747	4.604	7.173
5.	1.476	2.015	2.571	3.365	4.032	5.893
6.	1.440	1.943	2.447	3.143	3.707	5.208
7.	1.415	1.895	2.365	2.998	3.499	4.782
8.	1.397	1.860	2.306	2.896	3.355	4.499
9.	1.383	1.833	2.262	2.821	3.250	4.296
10.	1.372	1.812	2.228	2.764	3.169	4.143
11.	1.363	1.796	2.201	2.718	3.106	4.024
12.	1.356	1.782	2.179	2.681	3.055	3.929
13.	1.350	1.771	2.160	2.650	3.012	3.852
14.	1.345	1.761	2.145	2.624	2.977	3.787
15.	1.341	1.753	2.131	2.602	2.947	3.733
16.	1.337	1.746	2.120	2.583	2.921	3.686
17.	1.333	1.740	2.110	2.567	2.898	3.646
18.	1.330	1.734	2.101	2.552	2.878	3.610
19.	1.328	1.729	2.093	2.539	2.861	3.579
20.	1.325	1.725	2.086	2.528	2.845	3.552
21.	1.323	1.721	2.080	2.518	2.831	3.527
22.	1.321	1.717	2.074	2.508	2.819	3.505
23.	1.319	1.714	2.069	2.500	2.807	3.485
24.	1.318	1.711	2.064	2.492	2.797	3.467
25.	1.316	1.708	2.060	2.485	2.787	3.450
26.	1.315	1.706	2.056	2.479	2.779	3.435
27.	1.314	1.703	2.052	2.473	2.771	3.421
28.	1.313	1.701	2.048	2.467	2.763	3.408
29.	1.311	1.699	2.045	2.462	2.756	3.396
30.	1.310	1.697	2.042	2.457	2.750	3.385
35.	1.306	1.690	2.030	2.438	2.724	3.340
40.	1.303	1.684	2.021	2.423	2.704	3.307
45.	1.301	1.679	2.014	2.412	2.690	3.281
50.	1.299	1.676	2.009	2.403	2.678	3.261

Source: Engineering Statistics Handbook (National Bureau of Standards)

http://www.itl.nist.gov/div898/handbook/eda/section3/eda3672.htm

Answers for chapter questions.

One sample questions.

8- one sample-1. The problem tells us that the mean is $89,000, the number in the sample (n) = 9, the sample mean is $97,000, and the sample standard deviation is $10,000.

a. Ho: $\mu = \$89,000$

b. H1: $\mu > \$89,000$

c. The degrees of freedom will be n – 1. That is, 9-1 or 8. At the five percent level of significance and 8 degrees of freedom, the critical value of t is 1.860.

d. The decision rule would be: "Reject Ho if t > 1.860.

e.
$$t = \frac{\bar{X} - \mu}{S_{\bar{X}}}$$

$$t = \frac{\$97,000 - \$89,000}{\$10,000 / \sqrt{9}}$$

$$t = \frac{\$8,000}{\$10,000 / 3}$$

t = ($8,000 * 3) / $10,000

t = 2.400

f. Reject H_o

8 -one sample-2.

$\mu = \$75,000$
n = 3
X = $70,000
S = $5,000

a. Ho: $\mu = \$75,000$

b. H1: $\mu < \$75,000$

c. At 6 df (7-1) and five percent significance t = 1.943

d. Reject Ho if t < - 1.943

e.
$$t = \frac{\$70,000 - \$75,000}{\$5,000 / \sqrt{7}} = 2.646$$

f. Reject H_o.

8 — one sample-3.

$\mu = \$100,000$
n = 3
$\bar{X} = \$120,000$
S = $25,000

a. Ho: $\mu = \$100,000$

b. H1: µ >$100,000
c. At 2 df (3-1) and one percent significance $t = 6.965$
d. Reject Ho if t < - 6.965

e.
$$t = \frac{\$120,000 - \$100,000}{\$25,000 / \sqrt{3}} = 1.386$$

f. Do not reject H_o. The calculated value of t is not greater than the critical value from the table.

8-one sample-4.

$\mu = 97$
$n = 12$
$\overline{X} = 85$
$X = 85$
$S = 13$

a. H_o: $\mu = 97$
b. H1: $\mu < 97$
c. At 11 df (12-1) and one percent significance $t = 2.718$
d. Reject Ho if t < - 2.718

e.
$$t = \frac{85-97}{13 / \sqrt{12}} = -3.198$$

f. Reject H_o.

8-one sample-5.

$\mu = 27$
$n = 7$
$\overline{X} = 45$
$S = 17$

a. H_o: $\mu = 27$
b. H1: $\mu > 27$
c. At 6 df (7-1) and one percent significance $t = 3.143$
d. Reject Ho if t > 3.143

e.
$$t = \frac{45-27}{17 / \sqrt{7}} = 2.801$$

f. Do not reject H_o.

8- one sample-6. This is a two tail test. The one percent region of rejection will be divided between the right and left tails of the t distribution (one half percent in each tail). Therefore the critical value of t will be found in the table at the .005 significance level and six degrees of freedom.

$\mu = 27$
$n = 7$
$\bar{X} = 45$
$S = 17$

a. H_o: $\mu = 27$
b. $H1$: $\mu \neq 27$
c. At 6 df (7-1) and .005 significance $t = 3.707$
d. Reject H_o if $t > 3.707$ or < -3.707
e. The calculation of t is the same.

$$t = \frac{45-27}{17/\sqrt{7}} = 2.801$$

f. Do not reject H_o.

8- one sample-7.

$\mu = 5$
$n = 15$
$\bar{X} = 6.5$
$S = .5$

a. H_o: $\mu = 5$
b. $H1$: $\mu > 5$
c. At 14 df (15-1) and .05 significance $t = 1.761$
d. Reject H_o if $t > 1.761$

e. $$t = \frac{6.5-5}{.5/\sqrt{15}}$$

$$t = \frac{1.5}{.5/3.873}$$

$t = 1.5 (3.873 / .5)$

$t = 1.5 (7.746)$

$t = \underline{11.619}$

f. Reject H_o.

Two sample questions.

8- two sample-1. The problem gives us this information: $n_1 = 18$, $n_2 = 18$, $\bar{X_1}=21$, $\bar{X_2}=23$, $S_1 = 3$, and $S_2 = 4$.

a. Ho: $\mu 1 = \mu 2$

b. $H_1: \mu_1 \neq \mu_2$

c. $n1 + n2 - 2 = (18 + 18 - 2) = 34$ degrees of freedom: **t** = 2.030 (Using 35 degrees of freedom, the closest value on the t chart. Note: half of the region of rejection (.025) is in each tail.)

d. Reject H_o if **t** > 2.030 or < - 2.030.

e. The first step in calculating the value of the test statistic is to calculate the pooled variance.

$$S_p^2 = \frac{(n_1 - 1)(S_1^2) + (n_2 - 1)(S_2^2)}{n_1 + n_2 - 2}$$

$$S_p^2 = \frac{(18-1)(3^2) + (18-1)(4^2)}{18 + 18 - 2}$$

$$S_p^2 = \frac{(17)(9) + (17)(16)}{34}$$

$$S_p^2 = \frac{153 + 272}{34} = 12.5$$

Now that we have the pooled variance we can plug it into the formula to calculate the value of the test statistic.

$$t = \frac{\overline{X}_1 - \overline{X}_2}{\sqrt{S_p^2(1/n_1 + 1/n_2)}}$$

$$t = \frac{21 - 23}{\sqrt{12.5(1/18 + 1/18)}}$$

$$t = \frac{-2}{\sqrt{12.5(.1111)}}$$

$$t = \frac{-2}{\sqrt{1.389}}$$

$$t = \frac{-2}{1.1786} = -1.697$$

f. Since – 1.697 is not less than - 2.030, do not reject H_o.

8-two sample-2. The problem gives us this information: $n_1 = 7$, $n_2 = 9$, $\overline{X_1}=79$, $\overline{X_2}=83$, $S_1 = 15$, and $S_2 = 25$.

a. $H_o: \mu_1 = \mu_2$

b. $H_1: \mu_1 \neq \mu_2$

c. $n1 + n2 - 2 = (9 + 7 - 2) = 14$ degrees of freedom: **t** = 2.145 (Using .025 one tail value, because the five percent region of rejection is half in each tail.)

d. Reject Ho if **t** > 2.145 or < - 2.145.

The first step in calculating the value of the test statistic is to calculate the pooled variance.

$$S_p^2 = \frac{(n_1 - 1)(S_1^2) + (n_2 - 1)(S_2^2)}{n_1 + n_2 - 2}$$

$$S_p^2 = \frac{(7-1)(15^2) + (9-1)(25^2)}{7+9-2}$$

$$S_p^2 = \frac{(6)(225) + (8)(625)}{14}$$

$$S_p^2 = \frac{1{,}350 + 5{,}000}{14} = 453.5714$$

Now that we have the pooled variance we can plug it into the formula to calculate the value of the test statistic.

$$t = \frac{\overline{X}_1 - \overline{X}_2}{\sqrt{S_p^2(1/n_1 + 1/n_2)}}$$

$$t = \frac{79 - 83}{\sqrt{453.5714(1/7 + 1/9)}}$$

$$t = \frac{-4}{\sqrt{453.5714(.1429 + .1111)}}$$

$$t = \frac{-4}{\sqrt{115.207}}$$

$$t = \frac{-4}{10.733} = -.373$$

f. Because -.373 is not less than -2.145, do not reject H_o.

8-two sample-3. The problem provides:
$n_1 = 3$, $n_2 = 3$, $\overline{X}_1 = \$100{,}000$, $\overline{X}_2 = \$120{,}000$, $S_1 = 1{,}000$, and $S_2 = 1{,}000$.

a. $H_o: \mu_1 = \mu_2$

b. $H_1: \mu_1 < \mu$
c. $n1 + n2 - 2 = (3 + 3 - 2) = 4$ degrees of freedom: $t = 2.132$
d. Reject H_o if $t < -2.132$.

The first step in calculating the value of the test statistic is to calculate the pooled variance.

$$S_p^2 = \frac{(n_1 - 1)(S_1^2) + (n_2 - 1)(S_2^2)}{n_1 + n_2 - 2}$$

$$S_p^2 = \frac{(3-1)(1{,}000^2) + (3-1)(1{,}000^2)}{3 + 3 - 2}$$

$$S_p^2 = \frac{(2)(1{,}000{,}000) + (2)(1{,}000{,}000)}{4}$$

$$S_p^2 = \frac{4{,}000{,}000}{4} = 1{,}000{,}000$$

Now that we have the pooled variance we can plug it into the formula to calculate the value of the test statistic.

$$t = \frac{\overline{X}_1 - \overline{X}_2}{\sqrt{S_p^2 (1/n_1 + 1/n_2)}}$$

$$t = \frac{\$100{,}000 - \$120{,}000}{\sqrt{1{,}000{,}000(1/3 + 1/3)}}$$

$$t = \frac{\$ - 20{,}000}{\sqrt{\$1{,}000{,}000(.3333 + .3333)}}$$

$$t = \frac{\$ -20{,}000}{\sqrt{666{,}600}}$$

$$t = \frac{\$ -20{,}000}{816.456} = -24.496$$

f. Do not reject H_o.

8 -two sample -4. The problem gives us this information:

$n_1 = 29$, $n_2 = 12$, $\overline{X}_1 = 97$, $\overline{X}_2 = 85$, $S_1 = 10$, and $S_2 = 13$.

a. Ho: $\mu_1 = \mu_2$

b. H_1: $\mu_1 > \mu_2$

c. At one percent significance and **n1** + **n2** - 2 = (29 + 12 - 2) = 39 degrees of freedom: **t** = 2.423 (using the **t** table at **n** = 40, the closest value)
d. Reject Ho if **t** > 2.423.

The first step in calculating the value of the test statistic is to calculate the pooled variance.

$$S_p^2 = \frac{(n_1 - 1)(S_1^2) + (n_2 - 1)(S_2^2)}{n_1 + n_2 - 2}$$

$$S_p^2 = \frac{(29 - 1)(10^2) + (12 - 1)(13^2)}{29 + 12 - 2}$$

$$S_p^2 = \frac{28(100) + 11(169)}{39}$$

$$S_p^2 = \frac{4{,}659}{39} = 119.4615$$

Now that we have the pooled variance we can plug it into the formula to calculate the value of the test statistic.

$$t = \frac{\overline{X_1} - \overline{X_2}}{\sqrt{S_p^2(1/n_1 + 1/n_2)}}$$

$$t = \frac{97 - 85}{\sqrt{119.4615(1/29 + 1/12)}}$$

$$t = \frac{12}{\sqrt{119.4615(.0345 + .0833)}}$$

$$t = \frac{12}{\sqrt{14.0367}}$$

$$t = \frac{12}{3.751} = 3.199$$

f. Reject H_o.

Paired Differences (Dependent Samples)

8 -paired differences-1.

a. H_o: $\mu_d = 0$

b. H1: $\mu_d > 0$

c. The number of pairs minus one = 3 degrees of freedom. At the one percent level of significance the critical value of **t** = 4.541

d. Reject H_o if **t** > 4.541 (Note: If we subtract after values from before values this will result primarily in positive values. Therefore the average value of the differences (d) will be positive, and when this positive value is used in the formula to calculate **t**, the result will be a positive value of **t**.)

Intersection		1	2	3	4
Number before change		2	6	7	3
Number after change		3	3	2	1
	d	-1	3	5	2
	d²	1	9	25	4

The sum of the differences is 9, the average difference is 9/4 or 2.25, and the sum of the differences squared is 39.

e.
$$S_d = \sqrt{\frac{\Sigma d^2 - (\Sigma d)^2 / n}{n - 1}}$$

$$S_d = \sqrt{\frac{39 - (9)^2 / 4}{4 - 1}}$$

$$S_d = \sqrt{\frac{39 - 81/4}{4 - 1}}$$

$$S_d = \sqrt{\frac{39 - 20.25}{3}}$$

$$S_d = \sqrt{6.25}$$

$S_d =$ __2.5__

This value for the standard deviation of the differences can now be used to calculate the value of the t statistic.

$$t = \frac{\bar{d}}{S_d / \sqrt{n}}$$

$$t = \frac{2.25}{2.5 / \sqrt{4}}$$

t = (2.25 * 2) / 2.5 = 1.800

Since 1.800 is not greater than the critical value of the test statistic (4.541) the null hypothesis is not rejected.

8- paired differences-2.

a. H_o: $\mu_d = 0$

b. H_1: $\mu_d < 0$

c. The number of pairs minus one = 5 degrees of freedom. At the one percent level of significance the critical value of $t = 3.365$.

d. Reject H_o if $t < -3.365$

(Note: If we subtract after values from before values this will result primarily in negative values. Therefore the average value of the differences (d) will be negative, and when this negative value is used in the formula to calculate t, the result will be a negative value of t.)

Player	Able	Baker	Charlie	Dobson	Emmerson	Finegold
Interceptions last year	2	3	5	2	1	1
Interceptions this year	3	5	4	2	3	2
d	-1	-2	1	0	-2	-1
d^2	1	4	1	0	4	1

The sum of the differences is – 5, the average difference is -5/6 or - .8333, and the sum of the squares of the differences is 11.

e.

$$S_d = \sqrt{\frac{\Sigma d^2 - (\Sigma d)^2 / n}{n - 1}}$$

$$S_d = \sqrt{\frac{11 - (5)^2 / 6}{6 - 1}}$$

$$S_d = \sqrt{\frac{11 - 25/6}{5}}$$

$$S_d = \sqrt{\frac{11 - 4.1667}{5}}$$

$$S_d = \sqrt{\frac{6.833}{5}}$$

$$S_d = \sqrt{1.3667} = \underline{1.1690}$$

This value for the standard deviation of the differences can now be used to calculate the value of the t statistic.

$$t = \frac{\bar{d}}{S_d / \sqrt{n}}$$

$$t = \frac{-.8333}{1.1690 / \sqrt{6}}$$

$t = - .8333 * 2.4494 / 1.169$

$t = - 1.746$

Since - 1.746 is not greater than the critical value of the test statistic (3.365) the null hypothesis is not rejected.

8 –paired differences- 3.

Vine	Redbird Early	Mammoth Mid-season	Big Beef	Linger Longer	
Yield (old)	4.2	5.0	4.9	3.9	
Yield (new)	4.3	5.3	5.4	3.7	
d	-.1	-.3	-.5	.2	$\sum d = -.7$
d^2	.01	.09	.25	.04	$\sum d^2 = .39$

a. H_o: $\mu_d = 0$ $\bar{d} = -.7 / 4 = - 0.175$

b. H_1: $\mu_d < 0$

c. At 5 % and 3 df, critical value of **t** is 2.353

d. Reject H_o if t < - 2.353

e.
$$S_d = \sqrt{\frac{\sum d^2 - (\sum d)^2 / n}{n - 1}}$$

$$S_d = \sqrt{\frac{.39 - (-.7)^2 / 4}{4 - 1}} = .2986$$

$$t = \frac{\bar{d}}{S_d / \sqrt{n}}$$

$$t = \frac{-.175}{.2986 / \sqrt{4}}$$

$$t = \frac{-.175}{.2986 / 2} = - 1.172$$

f. Do not reject H_o.

8-paired diferences-4.

Month	Sept	Oct	Nov	Dec	Jan	Feb	Mar	Apr	May
Assaults (2007)	5	4	3	3	2	5	5	4	2
Assaults (2008)	2	3	2	4	4	2	3	1	0
d	-3	-1	-1	1	2	-3	-2	-3	-2
d^2	9	1	1	1	4	9	4	9	4

$\Sigma d = -12 \quad \Sigma d^2 = 42$

Note: The top row has been subtracted from the bottom row resulting in primarily negative differences and an alternate hypothesis test that says "less than zero". Had the bottom row been subtracted from the top row, resulting in predominantly positive differences, the alternate hypothesis would have been pointed in a positive direction. Either approach would be correct, as long as it is consistent with placing a negative or positive sign on the critical value of the test statistic from the table.

a. $H_o: M_d = 0$
b. $H_1: M_d < 0$
c. At one percent significance and eight df the critical value of t is -2.896.
d. Reject H_o if t < -2.896
e.
$$S_d = \sqrt{\frac{\Sigma d^2 - (\Sigma d)^2 / n}{n - 1}}$$

$$S_d = \sqrt{\frac{42 - (12)^2 / 9}{9 - 1}}$$

$S_d = 1.803$

$$t = \frac{\bar{d}}{S_d / \sqrt{n}}$$

$$t = \frac{-1.3333}{1.803 / 3} = -2.218$$

f. Do not reject H_o.

Chapter 9

Hypothesis tests using analysis of variance

What this chapter will do for you.

A researcher could use the hypothesis testing techniques we have looked at so far, in the previous chapters, to do one of two thing: (1) to compare a sample mean with a historical or expected population mean to see if the population mean has apparently changed or is not what it was thought to be, or (2) to compare samples from two populations to see if they have different means. Analysis of variance, often abbreviated as ANOVA ("ah-no'-va"), offers you a new capability. It can be used to compare three, four, or more populations, to determine if you should believe that their means are or are not all the same. Although explanations of ANOVA encountered in a typical statistics text may appear very confusing, the theory is simple and logical, and the procedure for making the calculations is not as complex as it may first appear. In this chapter you will learn the basic idea behind analysis of variance, and walk through the procedure for making the necessary calculations to reject, or fail to reject, the null hypothesis that the means of three or more populations are the same.

The "F" distribution.

The statistical distribution used for analysis of variance is called by mathematicians the "F" distribution. Unlike those we have looked at so far, the F distribution has only positive values, and one tail, which extends out to the right.

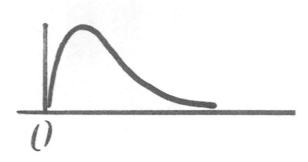

The F distribution begins at zero…and extends out to infinity.

There is a family of F distributions, depending on the number of "degrees of freedom" used in a particular problem. The structure of the problem will determine what that number will be.

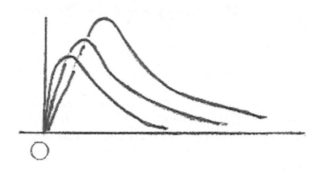

What does the F curve show? It shows, by the height of the curve, the frequency of accidental variations between sample values that would occur if we took samples from a number of <u>identical</u> populations (In analysis of variance these different populations are called "treatments"). Even if there were no difference at all between the means of treatments, and we took samples from each, the samples would always look different, just because sampling works that way. There is always some amount of accidental difference between samples (sampling error). Notice in the F curve shown above that the curve starts at zero. This means, in theory, that there would never be an instance, when samples were taken, that there would be no error at all. Some error would always be present. It would usually be a small error, but it would be present. A zero level of error (a value of zero for F) does not exist, but very small accidental errors are common however. We see this as we read the graph from zero error (the frequency curve at zero) to the right; the curve climbs rapidly and reaches a peak at small values of F. This shows that, even if the samples taken were from identical populations, small accidental variations would occur with great frequency.

As we move farther to the right, to larger F values that would correspond to greater accidental variation due to sampling, we notice that the curve drops rapidly and then tails off to the right. This indicates that, when there is really no difference between the treatments, large apparent differences due to sampling error become few in number. That is, there is a diminishing probability, in any one hypothesis test, that such an accidental apparent difference will occur if there is really no difference between the populations.

One way analysis of variance can look complicated, and the calculations can be burdensome for large problems. This is why personal computer programs are often used for ANOVA calculations. In fact, analysts sometimes use computer programs for ANOVA, without really knowing what the computer program is doing. It is important that you understand the theory behind the technique for two reasons: First, you might actually be required to conduct this kind of analysis. Second, and more likely, you may be in a business situation where you are the decision maker or are expected to advise the decision maker who is receiving a presentation on a hypothesis test carried out using analysis of variance. In the latter situation it would be important that you know something about the technique to better understand the merits of the conclusions being presented and possibly to know what questions to ask.

What is the theory behind this hypothesis test?

In general terms, calculation of the F statistics will provide a comparison of the variability between treatments, and the variability that exists within treatments.

In general terms:

$$F = \frac{\text{Variability between treatments}}{\text{Variability within treatments}}$$

When the ratio of observed variability between samples is large, compared to the observed variability within samples, the calculated F value (as suggested by the fraction above) will be large. This would place it far out in the right tail of the F distribution. Such a large value for the F statistic would be unlikely to occur by accident, through sampling error, if the means of the treatments were really all the same. So a large value for the calculated value of F suggests that the treatments are actually different.

Accidental variations between samples are more likely to occur if there is a great deal of variability in the size, weight, etc. of items within a population. Accidental variations are less likely if items in the populations are relatively uniform in size, weight, etc. If there is a great deal of variability within a population, as illustrated below, a sample could pick up a number of small items, or a number of large items. This would give a misleading picture of the actual average size of the items in the population.

This is an illustration of a population where the items in the population vary considerably in size.

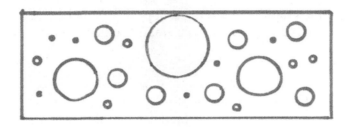

If a sample were taken from the population pictured above, it is entirely possible that it could contain a disproportionate number of large variables, or of small variables. In either case, the sample mean would not be representative of the population mean.

On the other hand, if items within a population are relatively uniform in size, as illustrated below, there would be little chance of a sample picking up abnormally large or abnormally small items, and the sample mean would be a fairly accurate representation of the population mean. It would be unlikely that such a sample would contribute to a large accidental variation between samples.

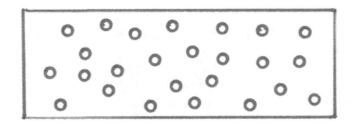

We cannot know for sure what the variation is item size within the populations is like, but we can get a good idea by measuring the variation in item size within the samples. A large amount of variability within samples would suggest that the samples were taken from populations where the items varied greatly in size. We would expect that variability to give rise to greater accidental differences between samples. It is logical then, that if the observed variability between samples is large relative to the variability within samples, then it is likely that there is a real difference between populations. But if the observed variability within samples is large, it is more likely that an apparent difference between samples is due simply to sampling error.

The null hypothesis and the alternate hypothesis.

For an ANOVA problem you will follow a procedure similar to that in setting up a Z or **t** test. You will need a null hypothesis and an alternate hypothesis. In this case the assumption stated in the null hypothesis is that the means of the populations being sampled are all equal.

$H_0 : \mu_1 = \mu_2 = \mu_3$

(If a fourth population is sampled, the null hypothesis would include μ_4, and so on.

The alternate hypothesis is that this is not the case. It is stated in words.

H1 : All the means are not equal.

To work an ANOVA problem it is necessary to be familiar with some terms peculiar to this kind of hypothesis test. ANOVA originated with analysts in the agricultural research business, who planted various test plots of crops and applied different treatments to them to judge the impact of these treatments on yields. The samples have traditionally been called "treatments". In the null hypothesis above (and in the problem which follows) the number of treatments is three. In the equations used for calculating the value of the test statistic, the number of treatments is symbolized by the letter K.

Setting up an ANOVA table.

To set up an ANOVA problem, the variables are written down in a table. This will be done in an example below. All the variables for the first treatment are written in a column, the variables for the second treatment are written in another column, and so forth, in as many columns as there are treatments in the problem. Spaces are provided just to the right of the variables in each column, so each variable can be squared and the value of the square written to the right of that variable. The number of variables taken in a sample from each treatment is symbolized by a lower case **n**, and since these variables are written down in columns - one column per treatment - the number of variables in column number one is called **n1**, the number of variables in column two is called **n2**, and so on. The total number of variables is symbolized by a capital N, which will be the sum of **n1**, **n2**, and so on.

To begin the calculations you will add up the variables in each column. These column totals will be written at the bottom of the column, and these totals will be arranged in a row at the bottom of the table. These column totals will be added across the page and the total written at the right. This will be the sum of all the variables (the X's), symbolized by ΣX.

The number of variables in each column will be written down. This row will be labeled nc, standing for column totals. The values in the nc row will be added across to the right and the sum written down. The will be the number of all the variables in the problem (N).

You would then square each variable in the problem, and write these squares just to the right of its variable. The resulting columns of squares are added up, and the results written in a row across the bottom of the table. When these values are added up at the right of the table the resulting number will be the summation of all the squares of the variables. This will be labeled ΣX^2.

Once you have completed the table, you will have the numbers needed to calculate the value of the test statistic.

The decision rule.

To set up a decision rule it is necessary to find the critical value of the test statistic (F) in a table of F values. Use of the F table may seem at first to be confusing, but you will find that it is actually not difficult. Table values are arranged by 'degrees of freedom'. Across the top of an F table you will find the words "Number of degrees of freedom in the numerator". This number of degrees of freedom is equal to the number of treatments minus one.

Why is it called "the number of degrees of freedom in the numerator"? "K" is the symbol for the number of treatments in a problem. The term "K – 1" appears in the numerator of the equation shown below, which will be used to calculate the value of F. In the sample problem below, if we subtract one from three we find that the number of degrees of freedom in the numerator is two.

Along the left side of the F table you will find the words "Number of degrees of freedom in the denominator". The number of degrees of freedom in the denominator, as used on the side of the F table, is the total number of variables in the problem minus the number of treatments, or "N - K". In the example below, with three treatments and 13 variables, N – K will be 13 minus 3. In this particular problem, the number of degrees of freedom in the denominator, "N – K", is ten.

Two F tables are commonly used: one for a five percent level of significance and one for a one percent level of significance. In working a problem you would choose the table that corresponds to the level of significance desired for the particular problem. Then, by picking the correct column heading (K-1) and following that column down until it intersects the row that is identified by N-K, you would find the critical value of the test statistic.

Using data in the problem table (the x's, squares of the x's, and so on) that you will have constructed, and inserting values into the formulas shown below, you would then calculate a value of F to compare with this critical value from the table.

Formulas used to calculate the F statistic.

Three formulas are used. The first two are used to calculate values to put into the third, which is then used to calculate the value for the test statistic (F). The first formula shown below is used to calculate what is called "the sum of the squares treatment". The term used is "SST".

$$SST = \Sigma \left(\frac{Tc^2}{n_c} \right) - \frac{(\Sigma X)^2}{N}$$

The second formula is used to calculate the "Sum of squares error".

$$SSE = \Sigma X^2 - \Sigma \left(\frac{Tc^2}{n_c} \right)$$

The third formula is used to calculate the value of the F statistic. It is a precise mathematical statement of what we said earlier was being measured in analysis of variance: the variability between samples divided by the variability within samples.

$$F = \frac{\frac{SST}{K-1}}{\frac{SSE}{N-K}}$$

Working an example.

Let us suppose that you are the owner of a large farm near Xenia, Ohio. To combat water runoff and provide water for your herds of cattle you have constructed three ponds of about two acres in size. To provide a recreational benefit you stocked each pond with 1,000 large-mouth bass and large numbers of bluegills and forage fish (minnows). Having read that fertilizing ponds increases the growth of plankton and other microscopic life, which then contributes to the growth of forage fish, and, in turn, the growth of the bass, you have fertilized the ponds regularly with equivalent amounts of three different kinds of fertilizer. You wish to see if all three varieties of fertilizer are equally effective. You obtain a sample of five fish from pond number one, four fish from pond number two, and four fish from pond number three so you can perform an analysis of variance.

You enter the weights of these fish in pounds in a table as shown below. Each fish weight is a variable (an X).

	Pond Number One		Pond Number Two		Pond Number Three		
	X	X²	X	X²	X	X²	
	10	------	11	------	8	------	
	20	------	9	------	10	------	
	15	------	11	------	11	------	
	10	------	10	------	12	------	
	20	------					Totals
T_c	____		____		____		$\Sigma X =$ ____
n_c	____		____		____		$N =$ ____
X^2	____		____		____		$\Sigma X^2 =$ ____

Following statement of the null hypothesis, the alternate hypothesis, and the decision rule, we will complete this table.

H₀: $\mu_1 = \mu_2 = \mu_3$

H₁: Not all of the means are equal.

For the decision rule we will need to go to an F table to obtain the critical value of the test statistic. Since 5 % is a commonly used lever of significance, we will use that table. The number of treatments is three so the number of degrees of freedom in the numerator will be 3 – 1 = 2. This indicates that we should use the second column of the table. The total number of variables is 5 + 4 + 4 = 13, so the number of degrees of freedom in the denominator would be 13 – 3 or ten.

This indicates that we should use the tenth line in the table. When we read down the second column to line ten, we find the critical value of the test statistic. It is 4.10. This means that, for the degrees of freedom we have in this problem, there is a probability of .05 (five percent) or less that the F statistic will calculate out to be 4.10 or larger, if there is really no difference in the mean weight of the fish in the ponds.

Our decision rule will then be "Reject **H**o if **F** > 4.10".

In the table of sample variables as shown above, we will follow the procedure described earlier: we will square the variables, add the variables in the columns, add the squares in the columns, and add across the lines at the bottom of the chart and write their totals at the bottom right of the chart. These calculated numbers are shown in bold faced italics.

	Pond Number One		Pond Number Two		Pond Number Three		
	X	X²	X	X²	X	X²	
	10	**_100_**	11	**_121_**	8	**_64_**	
	20	**_400_**	9	**_81_**	10	**_100_**	
	15	**_225_**	11	**_121_**	11	**_121_**	
	10	**_100_**	10	**_100_**	12	**_144_**	
	20	**_400_**					Totals
T_c	**_75_**		**_41_**		**_41_**		ΣX = **_157_**
n_c	**_5_**		**_4_**		**_4_**		N = **_13_**
X²		**_1,125_**		**_423_**		**_533_**	ΣX² = **_2,081_**

The numbers in the table can now be plugged into the formulas to calculate the value of the test statistic based on the sample data. First we will calculate the sum of the squares treatment.

$$SST = \Sigma \left(\frac{T_c^2}{n_c} \right) - \frac{(\Sigma X)^2}{N}$$

$$SST = \left[\frac{T_1^2}{n_1} + \frac{T_2^2}{n_2} + \frac{T_3^2}{n_3} \right] - \frac{(\Sigma X)^2}{N}$$

$$SST = \left[\frac{75^2}{5} + \frac{41^2}{4} + \frac{41^2}{4} \right] - \frac{(157)^2}{13}$$

$$SST = \left[\frac{5,625}{5} + \frac{1,681}{4} + \frac{1,681}{4} \right] - \frac{24,649}{13}$$

$$SST = \left[(1,125 + 420.25 + 420.25) \right] - 1,896.0769$$

$$SST = 1,965.5 - 1,896.0769$$

$$SST = \underline{69.4231}$$

We have calculated the sum of the squares treatment. Next we will calculate the sum of the squares error. This will be simpler because one of the numbers we need (the summation of the column totals squared divided by the number in the column) has already been calculated above.

$$SSE = \Sigma X^2 - \Sigma \left(\frac{T_c^2}{n_c} \right)$$

$$SSE = 2{,}081 - 1965.5$$

$$SSE = \underline{115.5}$$

With these values in hand, we can now calculate the value of the test statistic.

$$F = \frac{\dfrac{SST}{K-1}}{\dfrac{SSE}{N-K}}$$

$$F = \frac{\dfrac{69.4231}{3-1}}{\dfrac{115.5}{13-3}}$$

$$F = \frac{34.7116}{11.15} = 3.113$$

We can now make a decision on the null hypothesis. Since the calculated value of the test statistic is less than 4.10 (the critical value of the test statistic obtained from the F table) the null hypothesis cannot be rejected at the five percent confidence level, and the alternate hypothesis – that all the means are not equal – cannot be accepted.

What have you learned and what comes next?

You have learned the basic theory behind analysis of variance and how to perform the calculations for a simple problem. Typically, the numbers can be too burdensome for easy manual calculation and analysis of variance (ANOVA) problems are then completed using a statistical program running on a personal computer. A presentation of data analysis using ANOVA can be intimidating for someone who does not really understand what is behind the numbers. You, however, should be better able to cope with such a presentation.

In the next chapter you will have the opportunity to become familiar with a hypothesis testing technique that uses nominal data. In that technique the calculations tend to be quite simple and amenable to hand calculation.

Questions for this chapter.

9-1. Agronomists at The Ohio State University have planted test plots to a new variety of sweet corn called "Syrup and Honey Sweet Corn". They wish to determine if there is any difference between test plots fertilized with commercial chemical fertilizer, residue from chicken houses, or residue from the cow barns. Yields in bushels per acre are given below. Conduct an Analysis of Variance (ANOVA) to determine if researchers should conclude, at the five percent significance level, that there is a difference in the yields.

Chemical Fertilizer	Chicken Fertilizer	Cow Fertilizer
50	60	50
40	50	40
60	60	50
40	60	60
30		

a. State the null hypothesis.
b. State the alternate hypothesis.
c. What would be the number of degrees of freedom in the numerator?
d. What would be the number of degrees of freedom in the denominator?
e. State the decision rule.
f. Calculate the value of the test statistic.
g. State whether the null hypothesis should be accepted or rejected.

9-2. Use the data from problem number one above but use a one percent level of significance.
a. State the null hypothesis.
b. State the alternate hypothesis.
c. What would be the number of degrees of freedom in the numerator?
d. What would be the number of degrees of freedom in the denominator?
e. State the decision rule.
f. Calculate the value of the test statistic.
g. State whether the null hypothesis should be accepted or rejected.

9-3. John Jones has been a chicken farmer for years. This year however, he has made a considerable capital investment in his farm and, using water from a river that runs through his property, has built four huge fish ponds to raise catfish for the commercial market. He is experimenting to determine if any of the four techniques for regulating his ponds give different catfish yields. (Fish are harvested three times per year.) Yields, in pounds per acre of pond area, are given in the table below. Determine at the five percent level if he should believe that there is a difference in yields.

a. State the null hypothesis.
b. State the alternate hypothesis.
c. What would be the number of degrees of freedom in the numerator?
d. What would be the number of degrees of freedom in the denominator?
e. State the decision rule.
f. Calculate the value of the test statistic.
g. State whether the null hypothesis should be accepted or rejected.

Technique # 1	Technique #2	Technique # 3	Technique # 4
100	150	140	100
110	140	140	120
100	120	100	100

9-4. In the following year the yields from ponds using techniques 1, 3, and 4 are the same, however the pond which used technique number two was destroyed when the river flooded and could not be used this year. Of the remaining three ponds, the yields were the same as last year. Was there a difference, at the five percent level, in yields?

a. State the null hypothesis.
b. State the alternate hypothesis.
c. What would be the number of degrees of freedom in the numerator?
d. What would be the number of degrees of freedom in the denominator?
e. State the decision rule.
f. Calculate the value of the test statistic.
g. State whether the null hypothesis should be accepted or rejected.

9-5. Carpenter's Chevrolet has four sales persons. The sales manager has been fired because he wrecked the "demonstrator" automobile he was driving. The new sales manager, Mr. Ruthless, wishes to determine if there is a difference in the sales performance of the four sales persons on the staff. He has kept records for the past four weeks. It appears that one salesperson's sales lag the others. If this is really the case, Mr. Ruthless would prefer to terminate him and bring in another sales person that he knows is looking for a job. However, Mr. Ruthless is experienced enough at the trade to know that, for various reasons, the production of individuals in sales can go up and down through no fault of their own, and this small sample of four weeks could be misleading. He decides he must be extremely confident (at the one percent level) that the populations of production for these four individuals are really different before advising the sales person who appears to be less productive to find another job. Sales figures (number of automobiles sold) for the four week period are shown below.

a. State the null hypothesis.
b. State the alternate hypothesis.
c. What would be the number of degrees of freedom in the numerator?
d. What would be the number of degrees of freedom in the denominator?
e. State the decision rule.
f. Calculate the value of the test statistic.
g. State whether the null hypothesis should be accepted or rejected.

Jones	Smith	Harrison	Brown	Green
4	5	4	2	4
3	5	5	1	2
5	3	6	3	5
4	4	4	0	5

9-6. Suppose another month has gone past, and rightly or wrongly, Brown has been terminated. The remaining four salespersons are still on the staff. Appearances would seem to indicate that Harrison is the most productive salesperson. However, Mr. Ruthless wishes to conduct another Analysis of Variance to determine if he should believe, at the one percent confidence level, if there is really a difference between the sales of the four remaining sales people.

a. State the null hypothesis.
b. State the alternate hypothesis.
c. What would be the number of degrees of freedom in the numerator?
d. What would be the number of degrees of freedom in the denominator?
e. State the decision rule.
f. Calculate the value of the test statistic.
g. State whether the null hypothesis should be accepted or rejected.

Upper critical values of the F distribution for numerator degrees of freedom (column headings) and denominator degrees of freedom (line designations) 5% significance level

	1	2	3	4	5	6	7	8	9	10
1	161.448	199.500	215.707	224.583	230.162	233.986	236.768	238.882	240.543	241.882
2	18.513	19.000	19.164	19.247	19.296	19.330	19.353	19.371	19.385	19.396
3	10.128	9.552	9.277	9.117	9.013	8.941	8.887	8.845	8.812	8.786
4	7.709	6.944	6.591	6.388	6.256	6.163	6.094	6.041	5.999	5.964
5	6.608	5.786	5.409	5.192	5.050	4.950	4.876	4.818	4.772	4.735
6	5.987	5.143	4.757	4.534	4.387	4.284	4.207	4.147	4.099	4.060
7	5.591	4.737	4.347	4.120	3.972	3.866	3.787	3.726	3.677	3.637
8	5.318	4.459	4.066	3.838	3.687	3.581	3.500	3.438	3.388	3.347
9	5.117	4.256	3.863	3.633	3.482	3.374	3.293	3.230	3.179	3.137
10	4.965	4.103	3.708	3.478	3.326	3.217	3.135	3.072	3.020	2.978
11	4.844	3.982	3.587	3.357	3.204	3.095	3.012	2.948	2.896	2.854
12	4.747	3.885	3.490	3.259	3.106	2.996	2.913	2.849	2.796	2.753
13	4.667	3.806	3.411	3.179	3.025	2.915	2.832	2.767	2.714	2.671
14	4.600	3.739	3.344	3.112	2.958	2.848	2.764	2.699	2.646	2.602
15	4.543	3.682	3.287	3.056	2.901	2.790	2.707	2.641	2.588	2.544
16	4.494	3.634	3.239	3.007	2.852	2.741	2.657	2.591	2.538	2.494
17	4.451	3.592	3.197	2.965	2.810	2.699	2.614	2.548	2.494	2.450
18	4.414	3.555	3.160	2.928	2.773	2.661	2.577	2.510	2.456	2.412
19	4.381	3.522	3.127	2.895	2.740	2.628	2.544	2.477	2.423	2.378
20	4.351	3.493	3.098	2.866	2.711	2.599	2.514	2.447	2.393	2.348
21	4.325	3.467	3.072	2.840	2.685	2.573	2.488	2.420	2.366	2.321
22	4.301	3.443	3.049	2.817	2.661	2.549	2.464	2.397	2.342	2.297
23	4.279	3.422	3.028	2.796	2.640	2.528	2.442	2.375	2.320	2.275
24	4.260	3.403	3.009	2.776	2.621	2.508	2.423	2.355	2.300	2.255
25	4.242	3.385	2.991	2.759	2.603	2.490	2.405	2.337	2.282	2.236
26	4.225	3.369	2.975	2.743	2.587	2.474	2.388	2.321	2.265	2.220
27	4.210	3.354	2.960	2.728	2.572	2.459	2.373	2.305	2.250	2.204
28	4.196	3.340	2.947	2.714	2.558	2.445	2.359	2.291	2.236	2.190
29	4.183	3.328	2.934	2.701	2.545	2.432	2.346	2.278	2.223	2.177
30	4.171	3.316	2.922	2.690	2.534	2.421	2.334	2.266	2.211	2.165
31	4.160	3.305	2.911	2.679	2.523	2.409	2.323	2.255	2.199	2.153
32	4.149	3.295	2.901	2.668	2.512	2.399	2.313	2.244	2.189	2.142
33	4.139	3.285	2.892	2.659	2.503	2.389	2.303	2.235	2.179	2.133
34	4.130	3.276	2.883	2.650	2.494	2.380	2.294	2.225	2.170	2.123
35	4.121	3.267	2.874	2.641	2.485	2.372	2.285	2.217	2.161	2.114
36	4.113	3.259	2.866	2.634	2.477	2.364	2.277	2.209	2.153	2.106
37	4.105	3.252	2.859	2.626	2.470	2.356	2.270	2.201	2.145	2.098
38	4.098	3.245	2.852	2.619	2.463	2.349	2.262	2.194	2.138	2.091
39	4.091	3.238	2.845	2.612	2.456	2.342	2.255	2.187	2.131	2.084
40	4.085	3.232	2.839	2.606	2.449	2.336	2.249	2.180	2.124	2.077
41	4.079	3.226	2.833	2.600	2.443	2.330	2.243	2.174	2.118	2.071
42	4.073	3.220	2.827	2.594	2.438	2.324	2.237	2.168	2.112	2.065
43	4.067	3.214	2.822	2.589	2.432	2.318	2.232	2.163	2.106	2.059
44	4.062	3.209	2.816	2.584	2.427	2.313	2.226	2.157	2.101	2.054
45	4.057	3.204	2.812	2.579	2.422	2.308	2.221	2.152	2.096	2.049
46	4.052	3.200	2.807	2.574	2.417	2.304	2.216	2.147	2.091	2.044
47	4.047	3.195	2.802	2.570	2.413	2.299	2.212	2.143	2.086	2.039
48	4.043	3.191	2.798	2.565	2.409	2.295	2.207	2.138	2.082	2.035
49	4.038	3.187	2.794	2.561	2.404	2.290	2.203	2.134	2.077	2.030
50	4.034	3.183	2.790	2.557	2.400	2.286	2.199	2.130	2.073	2.026

Source: Engineering Statistics Handbook (National Bureau of Standards)

http://www.itl.nist.gov/div898/handbook/eda/section3/eda3673.htm

Upper critical values of the F distribution for numerator degrees of freedom (column headings) and denominator degrees of freedom (line designations)

1% significance level

	1	2	3	4	5	6	7	8	9	10
1	4052.19	4999.52	5403.34	5624.62	5763.65	5858.97	5928.33	5981.10	6022.50	6055.85
2	98.502	99.000	99.166	99.249	99.300	99.333	99.356	99.374	99.388	99.399
3	34.116	30.816	29.457	28.710	28.237	27.911	27.672	27.489	27.345	27.229
4	21.198	18.000	16.694	15.977	15.522	15.207	14.976	14.799	14.659	14.546
5	16.258	13.274	12.060	11.392	10.967	10.672	10.456	10.289	10.158	10.051
6	13.745	10.925	9.780	9.148	8.746	8.466	8.260	8.102	7.976	7.874
7	12.246	9.547	8.451	7.847	7.460	7.191	6.993	6.840	6.719	6.620
8	11.259	8.649	7.591	7.006	6.632	6.371	6.178	6.029	5.911	5.814
9	10.561	8.022	6.992	6.422	6.057	5.802	5.613	5.467	5.351	5.257
10	10.044	7.559	6.552	5.994	5.636	5.386	5.200	5.057	4.942	4.849
11	9.646	7.206	6.217	5.668	5.316	5.069	4.886	4.744	4.632	4.539
12	9.330	6.927	5.953	5.412	5.064	4.821	4.640	4.499	4.388	4.296
13	9.074	6.701	5.739	5.205	4.862	4.620	4.441	4.302	4.191	4.100
14	8.862	6.515	5.564	5.035	4.695	4.456	4.278	4.140	4.030	3.939
15	8.683	6.359	5.417	4.893	4.556	4.318	4.142	4.004	3.895	3.805
16	8.531	6.226	5.292	4.773	4.437	4.202	4.026	3.890	3.780	3.691
17	8.400	6.112	5.185	4.669	4.336	4.102	3.927	3.791	3.682	3.593
18	8.285	6.013	5.092	4.579	4.248	4.015	3.841	3.705	3.597	3.508
19	8.185	5.926	5.010	4.500	4.171	3.939	3.765	3.631	3.523	3.434
20	8.096	5.849	4.938	4.431	4.103	3.871	3.699	3.564	3.457	3.368
21	8.017	5.780	4.874	4.369	4.042	3.812	3.640	3.506	3.398	3.310
22	7.945	5.719	4.817	4.313	3.988	3.758	3.587	3.453	3.346	3.258
23	7.881	5.664	4.765	4.264	3.939	3.710	3.539	3.406	3.299	3.211
24	7.823	5.614	4.718	4.218	3.895	3.667	3.496	3.363	3.256	3.168
25	7.770	5.568	4.675	4.177	3.855	3.627	3.457	3.324	3.217	3.129
26	7.721	5.526	4.637	4.140	3.818	3.591	3.421	3.288	3.182	3.094
27	7.677	5.488	4.601	4.106	3.785	3.558	3.388	3.256	3.149	3.062
28	7.636	5.453	4.568	4.074	3.754	3.528	3.358	3.226	3.120	3.032
29	7.598	5.420	4.538	4.045	3.725	3.499	3.330	3.198	3.092	3.005
30	7.562	5.390	4.510	4.018	3.699	3.473	3.305	3.173	3.067	2.979
31	7.530	5.362	4.484	3.993	3.675	3.449	3.281	3.149	3.043	2.955
32	7.499	5.336	4.459	3.969	3.652	3.427	3.258	3.127	3.021	2.934
33	7.471	5.312	4.437	3.948	3.630	3.406	3.238	3.106	3.000	2.913
34	7.444	5.289	4.416	3.927	3.611	3.386	3.218	3.087	2.981	2.894
35	7.419	5.268	4.396	3.908	3.592	3.368	3.200	3.069	2.963	2.876
36	7.396	5.248	4.377	3.890	3.574	3.351	3.183	3.052	2.946	2.859
37	7.373	5.229	4.360	3.873	3.558	3.334	3.167	3.036	2.930	2.843
38	7.353	5.211	4.343	3.858	3.542	3.319	3.152	3.021	2.915	2.828
39	7.333	5.194	4.327	3.843	3.528	3.305	3.137	3.006	2.901	2.814
40	7.314	5.179	4.313	3.828	3.514	3.291	3.124	2.993	2.888	2.801
41	7.296	5.163	4.299	3.815	3.501	3.278	3.111	2.980	2.875	2.788
42	7.280	5.149	4.285	3.802	3.488	3.266	3.099	2.968	2.863	2.776
43	7.264	5.136	4.273	3.790	3.476	3.254	3.087	2.957	2.851	2.764
44	7.248	5.123	4.261	3.778	3.465	3.243	3.076	2.946	2.840	2.754
45	7.234	5.110	4.249	3.767	3.454	3.232	3.066	2.935	2.830	2.743
46	7.220	5.099	4.238	3.757	3.444	3.222	3.056	2.925	2.820	2.733
47	7.207	5.087	4.228	3.747	3.434	3.213	3.046	2.916	2.811	2.724
48	7.194	5.077	4.218	3.737	3.425	3.204	3.037	2.907	2.802	2.715
49	7.182	5.066	4.208	3.728	3.416	3.195	3.028	2.898	2.793	2.706
50	7.171	5.057	4.199	3.720	3.408	3.186	3.020	2.890	2.785	2.698

Source: Engineering Data Handbook (National Bureau of Standards)

http://www.itl.nist.gov/div898/handbook/eda/section3/eda3673.htm

Answers for Questions in Chapter Nine

9-1.

Chemical Fertilizer		Chicken Fertilizer		Cow Fertilizer	
X	X^2	X	X^2	X	X^2
50	2500	60	3600	50	2500
40	1600	50	2500	40	1600
60	3600	60	3600	50	2500
40	1600	60	3600	60	3600
30	900				
220		230		200	$\Sigma X = 650$
	10,200		13,300		10,200 $\Sigma X^2 = 33,700$
$n_c = 5$		$n_c = 4$		$n_c = 4$	

$N = 13$

a. $\mu_1 = \mu_2 = \mu_3$
b. All the means are not equal.
c. $K - 1 = 3 - 1 = 2$
d. $N - K = 13 - 3 = 10$
e. Reject Ho if $F > 4.103$

f.
$$SST = \Sigma \left(\frac{T_c^2}{n_c} \right) - \frac{(\Sigma X)^2}{N}$$

$$SST = \left(\frac{T_1^2}{n_1} + \frac{T_2^2}{n_2} + \frac{T_3^2}{n_3} \right) - \frac{(\Sigma X)^2}{N}$$

$$SST = \left(\frac{220^2}{5} + \frac{230^2}{4} + \frac{200^2}{4} \right) - \frac{(650)^2}{13}$$

$$SST = \left(\frac{48{,}400}{5} + \frac{52{,}900}{4} + \frac{40{,}000}{4} \right) - \frac{422{,}500}{13}$$

SST = (9,680 + 13,225 + 10,000) -- 32,500
SST = 32,905 - 32,500

SST = 405

You have calculated the sum of the squares treatment. Next you will calculate the sum of the squares error.

$$SSE = \Sigma X^2 - \Sigma \left(\frac{T_c^2}{n_c} \right)$$

SSE = 33,700 -- 32,905

SSE = 795

With these values in hand, you can now calculate the value of the test statistic.

$$F = \frac{\frac{SST}{K-1}}{\frac{SSE}{N-K}}$$

$$F = \frac{\frac{405}{3-1}}{\frac{795}{13-3}}$$

$$F = \frac{202.5}{79.5} = 2.547$$

g. Since the calculated value of F (2.547) is not larger than 4.103, the null hypothesis cannot be rejected.

9-2.
a. Same as 9-1.
b. Same as 9-1.
c. Same as 9-1.
d. Same as 9.1.
e. Reject Ho if F > 7.559
f. Same as 11-1.
g. Since 2.547 is not larger than 7.559, Ho cannot be rejected.

9-3.

#1	X^2	#2	X^2	#3	X^2	#4	X^2
100	10,000	150	22,500	140	19,600	100	10,000
110	12,100	140	19,600	140	19,600	120	14,400
100	10,000	120	14,400	100	10,000	100	10,000
310	32,100	410	56,500	380	49,200	320	34,400

$\sum X = 1{,}420 \quad \sum X^2 = 172{,}200$

$$SST = \sum \left(\frac{T_c^2}{n_c} \right) - \frac{(\sum X)^2}{N}$$

$$SST = \left(\frac{T_1^2}{n_1} + \frac{T_2^2}{n_2} + \frac{T_3^2}{n_3} + \frac{T_4^2}{n_4} \right) - \frac{(\sum X)^2}{N}$$

$$SST = \left(\frac{310^2}{3} + \frac{410^2}{3} + \frac{380^2}{3} + \frac{320^2}{3} \right) - \frac{(1{,}420)^2}{12}$$

$$SST = \left(\frac{310^2}{3} + \frac{410^2}{3} + \frac{380^2}{3} + \frac{320^2}{3} \right) - \frac{(1{,}420)^2}{12}$$

$SST = (32{,}033.333 + 56{,}033.333 + 48{,}133.333 + 34{,}133.333) - 2{,}016{,}400 / 12$

$SST = (170{,}333.333) - 168{,}033.333$

$SST = \underline{2{,}299.999}$

$$SSE = \sum X^2 - \sum \left(\frac{T_c^2}{n_c} \right)$$

$SSE = 172{,}200 - 170{,}333.333$

$SSE = \underline{1{,}866.777}$

$$F = \frac{\frac{SST}{K-1}}{\frac{SSE}{N-K}}$$

$$F = \frac{\frac{2,299.999}{4-1}}{\frac{1,866.777}{12-4}}$$

$$F = \frac{766.666}{233.347}$$

$$F = \underline{2.386}$$

a. $\mu_1 = \mu_2 = \mu_3 = \mu_4$
b. All the means are not equal.
c. K - 1 = 4 - 1 = 3
d. N – K = 12 – 4 = 8
e. Reject Ho if F > 4.066..
f. SST = 2,299.999, SSE = 1,866.667, F = 3.286
g. Do not reject Ho. The calculated value of the F statistic does is not greater than the critical value of F from the table.

9-4
a. Same as 9-3.
b. Same as 9-3.
c. K – 1 = 3 – 1 = 2
d. N – K = 9 – 3 = 6
e. Reject Ho if F > 5.143
f. SST = 955.55, SSE = 1400.01, F = 2.048
g. Do not reject Ho.

9-5

a. $\mu_1 = \mu_2 = \mu_3 = \mu_4 = \mu_5$
b. All the means are not equal.
c. 5-1 = 4
d. 20 - 5 = 15
e. Reject Ho if F > 3.056
f. SST = 24.7, SSE = 19.5, F = 4.750
g. Reject Ho

9-6

a. $\mu_1 = \mu_2 = \mu_3 = \mu_4 = \mu_5$

b. All the means are not equal.
c. 3
d. 12
e. Reject Ho if F > 5.953
f. SST = 1.5, SSE = 13.5, F = .444
g. Do not reject Ho

Chapter 10

Hypothesis tests using Chi-Square

What this chapter will do for you.

In studying this chapter you will learn how to use a hypothesis test that is particularly useful for nominal data. It is the most commonly used of what are called the non-parametric or distribution-free tests. This class of tests does not depend for their accuracy on assumptions regarding the distributions of the parent populations from which samples are taken. It lends itself to solution of a number of problems that may be encountered in business situations. It is a hypothesis test that was much used in academic research three decades ago, when pocket calculators were unknown, and only a few large institutions had a huge apparatus called "the computer" that filled a large room, was attended by its own mystic priesthood, and could do complex and cumbersome calculations overnight if supplied the previous evening with boxes of punched cards that provided it with numbers to crunch. Researchers at that time needed a technique where calculations could be done by hand or with a mechanical desk calculator. In other words, Chi-square (the "i" is pronounced like the "i" in "kite") is relatively simple and the calculations can often be done with pencil and paper.

The Chi-square distribution.

At first glance, the Chi-square distribution looks somewhat like the F distribution, but the mathematics underlying it are different. Like the F distribution, its exact shape also depends on degrees of freedom: With very few degrees of freedom it looks much like the F distribution. As the number of degrees of freedom increase, it begins to resemble the normal curve. As we have noted with other distributions, the region of rejection will lie somewhere out in the tail of the distribution. With the Z and t distributions it could be in one or both tails. The F distribution had only one tail, which extended out to the right. The Chi-square distribution is similar. It has only one tail, which extends out to the right. The calculated value of Chi-square is a measure of the difference between expected values and observed values. The illustration below shows that, as the number of degrees of freedom increases, this calculated value must become larger and larger if it is to indicate that this difference is actually significant, and not just a result of sampling error.

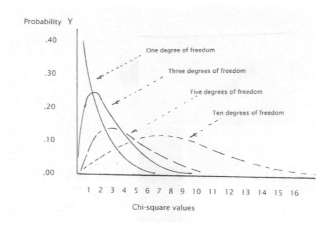

The theory behind Chi-square analysis.

Chi-square compares an observed set of data points with their "expected" values, and attempts to determine if there is something other than random chance involved in the differences between the two. Chi-square can be used for three types of analysis.

In the first, equal values are expected. An example would be a rotating wheel in a mechanical slot machine. Each of six sides, or faces, would be expected to have an equal chance of stopping behind the slot machine window. If the wheel were spun a large number of times, actual values could be compared with the expected values to determine whether or not the wheel is perfectly balanced. It is unlikely that such an experiment would result in equal appearances of each face; sampling error would insure some difference between equal and expected values. But accidental differences would usually be small. Chi-square analysis could be used to determine whether or not the differences were statistically significant.

In the second type of analysis, unequal values are expected. For example, the purchase decisions for motor vehicles in a particular region might be compared with the percentages for various kinds of vehicles purchased nationwide. If the percentages for purchases for sedans, minivans, and light pickup trucks were 40 %, 30 %, and 30 % nationwide, percentages for samples for particular sections of the country, such as the South, the Southwest, or New England could be compared to see if apparent differences are large enough to be statistically significant.

The most useful application of Chi-square analysis for business however, and the one that will be explained in detail here, is contingency table analysis, where two values are arrayed in a contingency table and a test conducted to determine if there is an association between the two variables. Chi-square will permit you to compare an observed outcome with the outcome that would be expected if only random chance were at work - that is, if there were no relationship between the variables. Small apparent differences could easily be due to chance, that is, to sampling error. Large differences however would suggest that there is an association between variables. What Chi-square can do for you is provide an analysis tool to evaluate the differences between expected and observed values, and determine, at a given level of confidence, whether or not it is logical to conclude that an association between the variables does, in fact, exist.

An example of a Chi-square analysis.

Rather than deal with words to further explain the concept, let's go directly to a sample problem to do so.

Suppose a civic club in County Seat, Indiana, has adopted as its primary objective the improvement of academic performance at its already-excellent local high school, and has suggested that students who work part time in businesses in the community may have lower grades because they are working when they should be studying. Business operators counter with the argument that students who work part time, on the average, are more serious students who not only work to help meet personal and family expenses but also apply themselves more diligently to their studies. A debate is raging in the community: Is part time employment associated with lower grades? Is it associated with higher grades? Or is there no association between part time work and academic performance?

The population in this case would consist of all the students at the high school, those who have, do, or will work part time, over a span of many years. Gathering information concerning past students would be difficult.

Information concerning future students is, of course, not yet available. Even gathering data for all the students presently in school might be burdensome and impractical. But we will assume that data for the senior class is actually available, and this class can be used as a sample to represent the entire population of past, present, and future students.

There are 100 students in the senior class. Exactly fifty of them are working part time while they attend school. Fifty are not. It just so happens that fifty of these 100 students are currently on the honor roll and fifty are not. These counts (nominal data) provide the material for a Chi-square test to determine whether or not residents should believe that there is an association between these two variables.

If there is no association between working and academic performance, it could be expected that any one working student, chosen at random, would have a fifty-fifty chance of being on the honor roll. The same would be true of students who did not work. Using these probabilities we will then construct the table of "expected" values found on the next page. Fifty students are in the category called "working". Since half of all students are on the honor roll, half (25 students) of the working students could be expected to be on the honor roll, and half (the other 25) not on it. Of the "not working" students in the second column, if only chance is involved in whether or not they are on the honor roll, then half (25) would be expected to be on it, and half (the other 25) not on it. (Note: The percentages would not necessarily be fifty-fifty. They might be 30-70, 25-75, or some other values.)

To put this another way: The ratios found in the column of totals at the right hand side of the table would be applied to each of the columns within the table. In the column of totals at the right hand side of the table, we see that fifty percent of all students are on the honor roll, and fifty percent are not. If there is no difference between working and non working students (if there is no association between the variables) this same ratio should hold for each of the other columns. Of the fifty working students, fifty percent of them (25 students) would be expected to be on the honor roll, and fifty percent (25 students) not on the honor roll. The same logic would be used to calculate the expected values for the non working column. If fifty percent of the total senior class of 100 is on the honor roll and fifty percent not, and there is no association between the variables, then fifty percent of the fifty non-working students (25 students) would be expected to be on the honor roll, and fifty percent (25 students) not on the honor roll.

Table of expected data

Student Category	Working - Expected Number	Not Working - Expected Number	(Totals)
On Honor Roll	25	25	50
Not on Honor Roll	25	25	50
(Totals)	50	50	100

In the next step in the Chi-square analysis, we obtain the actual values from our sample (the senior class) and discover, for the working group, that 30 students are on the honor roll, and 20 are not. We find out, for the group of students not working part time, that 20 are on the honor roll, and 30 are not. These figures have been entered in the contingency table below. For clarity, they are shown in italics.

Table of expected data and observed data

Student Category	Working Expected number	Working Observed number	Not Working Expected number	Not Working Observed number	(Totals) Exp.	(Totals) Actual
On Honor Roll	25	*30*	25	*20*	50	*50*
Not on Honor Roll	25	*20*	25	*30*	50	*50*
(Totals)	50	*50*	50	*50*	100	*100*

Calculating the value of Chi-square.

The numerical value for Chi-square is the summation, for each cell in the contingency table, of the following fraction: The numerator is the square of the quantity, observed value minus expected value, and the denominator is the expected value. This is shown algebraically on the next page.

$$X^2 = \frac{(f_o - f_e)^2}{f_e}$$

In this equation, f_o is the observed frequency, and f_e is the expected frequency.

In the upper left cell of the contingency table, for example, the value of Chi-square would be calculated as follows:

$$X^2 = \frac{(f_o - f_e)^2}{f_e}$$

$$X^2 = \frac{(30-25)^2}{25}$$

$$X^2 = \frac{5^2}{25}$$

$$X^2 = \frac{25}{25} = 1$$

The Chi-square calculation for the entire contingency table is simply the summation of the values for each cell of the table, calculated in the same manner as we have done for the first

$$X^2 = \frac{(f_o - f_e)^2}{f_e}$$

Next we will calculate the value of Chi-square for the four cells of the table.

$$X^2 = \frac{(30-25)^2}{25} + \frac{(20-25)^2}{25} + \frac{(20-25)^2}{25} + \frac{(30-25)^2}{25}$$

$$X^2 = \frac{(5)^2}{25} + \frac{(-5)^2}{25} + \frac{(-5)^2}{25} + \frac{(5)^2}{25}$$

$$X^2 = \frac{25}{25} + \frac{25}{25} + \frac{25}{25} + \frac{25}{25}$$

$$X^2 = 1 + 1 + 1 + 1 = 4$$

What does this number mean? To find out we need to go through the rest of the hypothesis testing steps. The null hypothesis is that there is no association between the variables. It is stated just like that – in words.

H_0: There is no association between the variables.

H_1: There is an association between the variables.

To find the critical value of the test statistic we would refer to a Chi-square table. To do so we must determine the number of degrees of freedom in the problem. This is not difficult. For a contingency table the degrees of freedom will be the number of rows in the table, minus one, multiplied times the number of columns in the table, minus one. For the table in the example this would be (2-1) x (2-1) = 1. We will then refer to the Chi-square table, using the column labeled .05 for five percent significance. (You could of course choose some other level of significance, such as one percent or ten percent, depending on how confident you wished to be if you rejected the null hypothesis.) The table shows us that the critical value of the test statistic is 3.841.

The decision rule then would be "Reject Ho if $X^2 > 3.841$

When we examined the table it appeared that working students tended to be on the honor roll with greater frequency that non working students. But that appearance could have been due to sampling error. The analysis shows however that there would be less than a five percent probability that there is no association between the variables. The null hypothesis is rejected and the business people, who insist that working students tend to do better in school, are triumphant.

What have you learned and what comes next?

You have learned how to use Chi-Square to analyze a contingency table, including the operation typically most difficult for student who encounter Chi-Square for the first time: calculating expected values. In the next chapter you will have the opportunity to become familiar with another hypothesis testing technique using nominal data –proportions.

Questions for this chapter.

10-1. The owner of the Buttermilk Diary has a herd if 100 prime milking cows. When he went to a lecture at the county agent's office he heard that soothing music played in the milking barn before and during milking increases yields. He is skeptical but decides to give it a try. Sixty Jersey cows are milked in barn one and forty Milking Shorthorn cows in barn two. For one month he tunes a radio to a soothing music station and plays it in both barns before and during milking. Then for one month the cows get no music. Yields in total pounds of milk, for the two months, are shown below. Determine at the five percent level of significance if one of the breeds of cows responds better to music.

a. State the null hypothesis.
b. State the alternate hypothesis.
c. What would be the number of degrees of freedom to be used in finding the critical value of the test statistic.
e. State the decision rule.
g. Calculate the value of the test statistic.
g. State whether the null hypothesis should be accepted or rejected.

	Jersey Cows	Milking Shorthorns
With Music	10,000	8,000
Without Music	9,000	7,000

10-2. A large number of children walking to school at the Happy Meadows Elementary School have had their lunch money stolen and have been harassed by older youth. People in the neighborhood attend a PTA meeting and agree to watch the sidewalks in their neighborhoods, call the police, and even confront youth committing these crimes – telling them the police are on their way. Crime data for last month (without neighborhood watch) is compared with crime for this month (with neighborhood watch). At the five percent significance level should we conclude that these petty crimes have been reduced in number?

	Neighborhood # One	Neighborhood # 2	Neighborhood # 3
Without	10	12	8
With	9	8	9

a. State the null hypothesis.
b. State the alternate hypothesis.
c. What would be the number of degrees of freedom to be used in finding the critical value of the test statistic.
e. State the decision rule.
f. Calculate the value of the test statistic.
g. State whether the null hypothesis should be accepted or rejected.

10-3. The coach of the basketball team at Western State College was not satisfied with the academic performance of the members of his squad during the preceding grading period. This period he is requiring them to attend study tables in the library, two hours per night following the evening meal. Using the data below should he conclude, at the five percent level, that there is an association between use of study tables and academic performance?

Possible grades	A	B	C	<C
Number receiving grades.				
Before tables	1	1	10	3
With tables	2	4	9	0

a. State the null hypothesis.
b. State the alternate hypothesis.
c. What would be the number of degrees of freedom to be used in finding the critical value of the test statistic.
e. State the decision rule.
f. Calculate the value of the test statistic.
g. State whether the null hypothesis should be accepted or rejected.

10-3. Sylvia Henderson depends on flower sales on two holidays during the year to provide the margin of profit for her florist shop. She wonders if advertising is more effective on one holiday than the other. She has sales figures for last year, when she did no advertising, and sales figures for this year, when she has advertised on local radio stations. She wonders if there is an association between the effectiveness of advertising and particular holidays. At the five percent significance level, should she believe this is the case?

Holiday	Valentine's Day	Mothers' Day
No Advertising	$50,000	$40,000
Advertising	$70,000	$50,000

a. State the null hypothesis.
b. State the alternate hypothesis.
c. What would be the number of degrees of freedom to be used in finding the critical value of the test statistic.
e. State the decision rule.
f. Calculate the value of the test statistic.
g. State whether the null hypothesis should be accepted or rejected.

10-5. A convenience food chain has stores in five neighborhoods. It decides to experiment with extended hours (open 24 hours versus closing at ten pm). It has been suggested that the impact of extended hours will be different for different stores. Sales increased at all five stores. Should management conclude, at the five percent level of significance, that there is an association between various stores and the impact of extended hours? Average daily sales, in thousands of dollars, are shown below.

	College Corner	Peaceful Hill	Midtown	Truckers' Quick Stop	Country Acres
10 pm	1	3	2	2	2
24 hr.	3	4	3	4	3

a. State the null hypothesis.
b. State the alternate hypothesis.
c. What would be the number of degrees of to be used in finding the critical value of the test statistic.

e. State the decision rule.
f. Calculate the value of the test statistic.
g. State whether the null hypothesis should be accepted or rejected.

Upper critical values of chi-square distribution with degrees of freedom listed in the column headings

	Probability of exceeding the critical value				
	0.10	0.05	0.025	0.01	0.001
1	2.706	3.841	5.024	6.635	10.828
2	4.605	5.991	7.378	9.210	13.816
3	6.251	7.815	9.348	11.345	16.266
4	7.779	9.488	11.143	13.277	18.467
5	9.236	11.070	12.833	15.086	20.515
6	10.645	12.592	14.449	16.812	22.458
7	12.017	14.067	16.013	18.475	24.322
8	13.362	15.507	17.535	20.090	26.125
9	14.684	16.919	19.023	21.666	27.877
10	15.987	18.307	20.483	23.209	29.588
11	17.275	19.675	21.920	24.725	31.264
12	18.549	21.026	23.337	26.217	32.910
13	19.812	22.362	24.736	27.688	34.528
14	21.064	23.685	26.119	29.141	36.123
15	22.307	24.996	27.488	30.578	37.697
16	23.542	26.296	28.845	32.000	39.252
17	24.769	27.587	30.191	33.409	40.790
18	25.989	28.869	31.526	34.805	42.312
19	27.204	30.144	32.852	36.191	43.820
20	28.412	31.410	34.170	37.566	45.315
21	29.615	32.671	35.479	38.932	46.797
22	30.813	33.924	36.781	40.289	48.268
23	32.007	35.172	38.076	41.638	49.728
24	33.196	36.415	39.364	42.980	51.179
25	34.382	37.652	40.646	44.314	52.620
26	35.563	38.885	41.923	45.642	54.052
27	36.741	40.113	43.195	46.963	55.476
28	37.916	41.337	44.461	48.278	56.892
29	39.087	42.557	45.722	49.588	58.301
30	40.256	43.773	46.979	50.892	59.703
35	46.059	49.802	53.203	57.342	66.619
40	51.805	55.758	59.342	63.691	73.402
45	57.505	61.656	65.410	69.957	80.077
50	63.167	67.505	71.420	76.154	86.661

Source: Engineering Data Handbook (National Bureau of Standards

http://www.itl.nist.gov/div898/handbook/eda/section3/eda3674.htm 12-13

Answers for Chapter Questions

10-1.
a. There is no association between the variables.
b. There is an association between the variables.
c. (2-1) X (2-1) = 1
e. Reject Ho if $X^2 > 3.841$
f. Calculations:

	Jersey Cows	Expected Value	Milking Shorthorns	Expected Value	Totals	Decimal
With Music	10,000	10,070*	8,000	7,950	18,000	.53
Without Music	9,000	8,930	7,000	7,050	16,000	.47
Totals	19,000	19,000	15,000	15,000	34,000	1.00

* 19,000 x .53

$$X^2 = \Sigma \left(\frac{(f_o - f_e)^2}{f_e} \right)$$

$$X^2 = \frac{(10,000-10,070)^2}{10,070} + \frac{(8,000-7,950)^2}{7,950} + \frac{(9,000-8,930)^2}{8,930} + \frac{(7,000-7,050)^2}{7,050}$$

$$X^2 = \frac{-70^2}{10,070} + \frac{50^2}{7,950} + \frac{70^2}{8,930} + \frac{-50^2}{7,050}$$

$$X^2 = \frac{4,900}{10,070} + \frac{2,500}{7,950} + \frac{4,900}{8,930} + \frac{2,500}{7,050}$$

$X^2 = 0.486 + 0.315 + 0.549 + 0.355 = \underline{1.705}$

g. Even though it appears that the milking shorthorn cows are more influenced by the music, the calculated value of the test statistic is less than the critical value of 3.841, found in the .05 significance column of the table, at one degree of freedom, and the null hypothesis is not rejected.

10-2.
a. There is no association between the variables.
b. There is an association between the variables.
c. (3-1) X (2-1) = 2
e. Reject Ho if $X^2 > 5.991$
f. Calculations:

	#1 f_o	f_e	#2 f_o	f_e	#3 f_o	f_e	Totals	Decimal
Without	10	10.26	12	10.80	8	9.18	30	.54
With	9	8.74	8	9.20	9	7.82	26	.46
Totals	19	19.00	20	20.00	17	17.00	56	1.00

X^2 = .007 + .133 + .142 + .008 + .157 + .178
X^2 = .625
g. The null hypothesis cannot be rejected.

10-3.
a. There is no association between the variables.
b. There is an association between the variables.
c. (4-1) X (2-1) = 3
e. Reject Ho if X^2 > 7.815
f. Calculations:

Possible grades	A	fe	B	fe	C	fe	< C	fe	Totals	Dec.
Number receiving grades.										
Before tables	1	1.5	1	2.5	10	9.5	3	1.5	15	.50
With tables	2	1.5	4	2.5	9	9.5	0	1.5	15	.50
Totals	3	3	5	5	19	19	3	3	30	1.00

X^2 = .167 + .9 + .026 + 1.5 + .167 + .9 + .026 + 1.5
X^2 = 5.186

g. Ho is not rejected.

10-4
a. There is no association between the variables.
b. There is an association between the variables.
c. (2-1) X (2-1) = 1
e. Reject Ho if X^2 > 3.841
f. Calculations:

Holiday Decimal	Valentine's Day		Mothers' Day		Totals	
	(fo)	fe	(fo)	fe		
No Advertising	$50,000	$51,600	$40,000	$38,700	$90,000	.43
Advertising	$70,000	$68,400	$50,000	$51,300	$120,000	.57
Totals	$120,000	$120,000	$90,000	$90,000	$210,000	1.00

X^2 = 49.612 + 43.669 + 37.430 + 30.018
X^2 = 160.729

g. Ho is rejected.

10-5.
a. There is no association between the variables.
b. There is an association between the variables.

c. (5-1) x (2-1) = 4
e. Reject Ho if $X^2 > 9.488$.
f. Calculations:

	College Corner		Peaceful Hill		Midtown		Truckers' Quick Stop		Country Acres		Totals	Dec.
		fe		*fe*		*fe*		*fe*		*fe*		
10 pm	1	1.48	3	2.59	2	1.85	2	2.22	2	1.85	10	.37
24 hr.	3	2.52	4	4.41	3	3.15	4	3.78	3	3.15	17	.63
	4	4	7	7	5	5	6	6	5	5	27	1.00

$X^2 = .5144$

g. The null hypothesis is not rejected.

Chapter 11

Hypothesis tests using Proportions

What this chapter will do for you.

In studying this chapter you will learn how to use another hypothesis test that is useful for nominal data. Proportions are arrived at by counting things. For example, an advertiser might make the claim: "When surveyed six months after making their purchase, ninety five of every 100 people who purchased a new Huffy Bicycle in 2009 rated their purchase as excellent." Or a nationwide survey might indicate that eight out of ten Americans who have a choice prefer to watch cable channels in preference to the over-the-air broadcast channels. Since ratio data is not required, analysis of proportions could be important in certain business situations. After mastering the material in this chapter, you will be able to carry out hypothesis tests involving proportions.

The distribution commonly used for these tests.

Numbers involved in proportion problems are typically large. Consequently, the distribution used in the problems in this chapter is the Z distribution. Depending on the problem, the region of rejection for the null hypothesis could be in the right tail or the left tail, if it is a one tail test, or it could be split between right and left tails of the distribution if the problem involves a two tail test.

Types of hypothesis tests involving proportions.

The letter P will stand for the proportion in the population. Two kinds of calculations are possible: In the first kind, a sample can be taken to see if a particular group differs in the proportion having some attribute from what the proportion was assumed to be. In this case the null hypothesis would be

H_0: P = some value. The alternate hypothesis could be H_1: P > some value,
H_1: P < some value, or H_1: P is not equal to some value.

In the second kind, two samples can be taken, one from each of two populations, to test the null hypothesis that the populations are the same with respect to some attribute. In this case the null hypothesis would be H_0: $P_1 = P_2$. Again, three alternate hypotheses are possible: H_1: $P_1 > P_2$,

H_1: $P_1 < P_2$, or H_1: P_1 is not equal to P_2.

A one sample test.

In these examples a capital letter P will stand for a population proportion. Just as X "X bar") stands for a sample mean, $\bar{P}$ ("P bar") will be the symbol for a sample proportion.

In our sample problem, let us suppose that Honda of America Manufacturing, Inc. (HAM) has reported that three out of ten purchasers of new Honda automobiles in 2004, in the first five years of ownership, reported no problems with their vehicles requiring more than routine maintenance. Helpful Honda, located in Dayton, Ohio is a dealership that has been in business many years and had sold thousands of automobiles in that time. The dealership prides itself on making sure that all cars sold are in perfect working order before delivery and owners are personally instructed in care and maintenance of their purchase. Consequently, it believes that an even greater proportion of its new car customers have had no problems. The dealership surveys 100 customers who purchased new Hondas from Helpful Honda and have had their cars a full five years. Forty of the 100 purchasers surveyed have had no problems with their cars. You are asked by the manager to conduct a statistical analysis, at the .05 significance level, to determine if she can conclude that a larger proportion of Helpful Honda customers have had no problems.

Setting up the problem.

We will first determine the null and alternate hypotheses.

$H_0: P = .3 \quad\quad H_1: P > .3$

Since this is a one tail test to be conducted at the five percent significance level we refer to the Z table and find the Z value where the area under the curve is 45 percent of all the variables in the distribution. The value of Z is 1.645.

The decision rule then is "Reject Ho if Z > 1.645"

The sample proportion is the number of "successes" in the sample (in this case 40) divided by the total number of respondents in the survey (100).

$$\bar{P} = \frac{40}{100}$$

The formula for calculating the Z value in a one sample test of proportions is stated as follows:

$$Z = \frac{\bar{P} - P}{\sqrt{\frac{P(1-P)}{n}}}$$

Substituting our problem values in the formula:

$$Z = \frac{.40 - .30}{\sqrt{\frac{.30(1-.30)}{100}}}$$

$$Z = \frac{.10}{\sqrt{\frac{.21}{100}}}$$

$$Z = \frac{.10}{\sqrt{.0021}}$$

$$Z = \frac{.10}{.0458} = 2.18$$

In our sample problem, since the calculated value of the test statistic (2.18) exceeds the critical value of the test statistic (1.645) the null hypothesis is rejected. The manager can conclude, at the .05 level of significance, that the autos sold to customers of Happy Honda are more trouble free than the national average.

You have probably already guessed that the square root term in the denominator of the fraction used to calculate the Z value is the standard error of the proportion – corresponding to the standard error of the mean in the Z problems you worked when you began hypothesis tests. The formula for a two sample test has a similar term.

A two sample test.

You may recall that in a two sample test of population means using ratio data, as carried out in Chapter Nine, the absolute difference between two sample means was divided by the standard deviation of the normal distribution of differences, plus and minus, that would occur by chance between two sample means when the two samples were taken from the same population, or from identical populations. In a similar manner, the absolute difference between two sample proportions is divided by the standard deviation of the normal distribution of differences, plus and minus, that would occur by chance between two sample proportions when the two samples were taken from the same population, or from identical populations. The formula for calculating the Z statistic in a two sample test of proportions is provided below.

$$Z = \frac{\overline{P_1} - \overline{P_2}}{\sqrt{\frac{P_c(1-P_c)}{n_1} + \frac{P_c(1-P_c)}{n_2}}}$$

$\overline{P_1}$ is the proportion in the first sample. $\overline{P_2}$ is the proportion in the second sample.
n_1 is the number in the first sample. n_2 is the number in the second sample.

P_c is the pooled proportion. It is found by adding the successes in the first and second sample, and dividing this sum by the sum of the numbers In the two samples.

$$P_c = \frac{X_1 + X_2}{n_1 + n_2}$$

X_1 is the number of successes in the first sample.
X_2 is the number of successes in the second sample.

Setting up the problem.

The Republic Department Store in Mount Pleasant, Michigan has attempted to improve its already high level of customer service. In a sample of 200 customers who made a purchase at the store in September 2008, 90 rated customer service as excellent. In a survey of 300 customers who purchased items six months later, in March 2009, 150 rated customer service as excellent. You have been asked to determine, at the .10 level of significance, if the store manager can conclude that there is a difference in perceptions of customer service between the two months.

All the customers who purchased items at the store in September 2008 would be the first population, and all the customers who purchased items at the store in March 2009 would be the second population. These two samples, of 200 customers and 300 customers respectively, are taken from populations that are much larger than the samples.

The null hypothesis - $H_0: p_1 = p_2$
The alternate hypothesis - $H_1: p_1 \neq p_2$

Because the alternate hypothesis simply says that there is a difference, the difference could go either way. Since a 10 percent significance level was specified, half of the 10 percent region of rejection (or five percent) would be in each tail of the Z distribution. Consequently, the decision rule would be written as follows:

"Reject H_0 if Z > 1.645 or < - 1.645"

The region of rejections would be as pictured below.

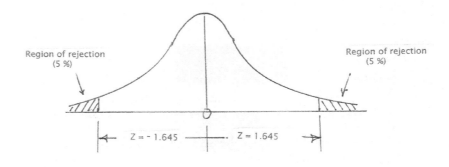

Calculating the value of the test statistic

The first step will be to find the pooled variance.

$$Pc = \frac{X_1 + X_2}{n_1 + n_2}$$

$$Pc = \frac{90 + 150}{200 + 300}$$

$$Pc = \frac{240}{500} = 0.48$$

$\overline{P}_1 = 90 / 200 \quad \overline{P}_2 = 150 / 300$

$\overline{P}_1 = .45 \quad \overline{P}_2 = .50$

Using the pooled proportion and the formula provided earlier we can now calculate the value of Z, the test statistic.

$$Z = \frac{\overline{P}_1 - \overline{P}_2}{\sqrt{\frac{Pc(1-Pc)}{n_1} + \frac{Pc(1-Pc)}{n_2}}}$$

$$Z = \frac{.45 - .50}{\sqrt{\frac{.48(1-.48)}{200} + \frac{.48(1-.48)}{300}}}$$

$$Z = \frac{-.05}{\sqrt{\frac{.2496}{200} + \frac{.2496}{300}}}$$

$$Z = \frac{-.05}{\sqrt{.001248 + .00083}}$$

$$Z = \frac{-.05}{\sqrt{.00208}} = -.05 / .0456 = -1.09649 \approx -1.10$$

Since 1.10 is neither smaller than – 1.645 nor larger than + 1.645 the null hypothesis is not rejected. There is not adequate evidence to conclude that there is now a difference in perceptions of customer service.

What have you learned and what comes next?

Proportions can be useful, particularly when only nominal data from counting something is available. The proportion of satisfied customers, the proportion of some good returned to stores, the proportion of flights arriving on time, and so forth may be valuable measures, and the means to determine if a statistically significant change has occurred because of some outside influence or because of some action management has taken. This chapter has provided an introduction to both one sample and two sample tests of proportions. In the next chapter you will leave the subject of hypothesis testing and study index numbers.

Questions for this chapter.

11-1. In December 2009 the Bureau of Labor Statistics reported that the nationwide unemployment rate was 10 %. The Dayton newspaper surveyed a random sample of 100 individuals of working age to find out if they were employed or unemployed and determined that fifteen were unemployed. At the five percent significance level does this indicate that the unemployment rate is higher in the Dayton area?
a. What is the null hypothesis?
b. What is the alternate hypothesis?
c. What is the critical value of the test statistic?
d. What is the decision rule?
e. Calculate the value of the test statistic.
f. Indicate your decision on the null hypothesis.

11-2. In 2009, forty percent of the graduates for all Michigan state-assisted colleges and universities had jobs by July Fourth of the year they graduated. You take a sample of two hundred 2009 graduates in business administration and discover that 95 had jobs. At the one percent significance level do a higher percentage of business administration graduates have jobs.
a. What is the null hypothesis?
b. What is the alternate hypothesis?
c. What is the critical value of the test statistic?
d. What is the decision rule?
e. Calculate the value of the test statistic.
f. Indicate your decision on the null hypothesis.

11-3 Suppose the five percent significance level had been chosen. Would it have made a difference in the decision on the null hypothesis?

11-4. A possible reform candidate for governor of a Midwestern state has been told that, to win, she must achieve at least seventy percent of the vote in downstate counties to overcome the deficit she will encounter in the state's big metropolitan area. Her committee surveys 400 likely voters in downstate counties and finds that 310 of them indicate they will vote for her. At the one percent significance level can she be confident that, if she runs, she will achieve her desired level of support?
a. What is the null hypothesis?
b. What is the alternate hypothesis?
c. What is the critical value of the test statistic?
d. What is the decision rule?
e. Calculate the value of the test statistic.
f. Indicate your decision on the null hypothesis.

11-5. The Gander Mountain Hospital System employs registered nurses that have four year degrees and nurses who graduated from three year nursing programs. There is a perception that nurses with degrees are more likely to be promoted to supervisory positions. As an intern in the CEO's office, has been assigned the task of conducting a hypothesis test at the ten percent level of significance to shed light on this situation. The CEO wishes to know if he should conclude that a higher proportion of degreed nurses are in supervision than non degreed nurses. Sample data is shown below.

	Sample size	Number in Supervisory positions
Four year degree	100	27
No degree	200	39

a. What is the null hypothesis?
b. What is the alternate hypothesis?
c. What is the critical value of the test statistic?
d. What is the decision rule?
e. Calculate the value of the test statistic.
f. Indicate your decision on the null hypothesis.

11-6. A problem has developed with the F-40 Fighter Jet, which has been in service for a year. Fatigue cracks have developed in the structural members holding the jet engine in place. This is an alarming problem that must be corrected. The maintenance officer is puzzled. A senior non-commissioned officer with long years of maintenance service suggests to the maintenance officer that it appears there is a difference between aircraft produced by the Warbird Corporation's Columbus factory and the company's Toledo factory. At the one percent significance level, does the data shown in the table below support his theory?

	Number of aircraft inspected	Number with problems
Columbus factory	35	10
Toledo factory	40	11

a. What is the null hypothesis?
b. What is the alternate hypothesis?
c. What is the critical value of the test statistic?
d. What is the decision rule?
e. Calculate the value of the test statistic.
f. Indicate your decision on the null hypothesis.

11-7. The Wash N' Press uniform service supplies uniforms to food service companies, health facilities, and other businesses in fourteen eastern states. When a truck requires unscheduled maintenance, it causes difficulties in scheduling pickup and delivery of uniforms at the commercial establishments they serve. They presently use two different brands of delivery trucks and are considering gradually going to only one brand as they replace vehicles over the next two years. The fleet manager has suggested that brand B trucks are particularly troublesome with respect to requirements for unscheduled maintenance and has recommended that data be collected for six months (120 working days) to compare proportions of each brand needing maintenance outside of that regularly scheduled.. At the five percent level, does the sample data shown below indicate that brand A is superior to brand B with respect to number of days out of service for unscheduled maintenance?

	Brand A trucks in fleet	Brand B trucks in fleet
	500	500
Number requiring unscheduled Maintenance	35	50

a. What is the null hypothesis?
b. What is the alternate hypothesis?
c. What is the critical value of the test statistic?
d. What is the decision rule?

e. Calculate the value of the test statistic.
f. Indicate your decision on the null hypothesis.

Area under the Normal Curve from 0 to X

X	0.00	0.01	0.02	0.03	0.04	0.05	0.06	0.07	0.08	0.09
0.0	0.00000	0.00399	0.00798	0.01197	0.01595	0.01994	0.02392	0.02790	0.03188	0.03586
0.1	0.03983	0.04380	0.04776	0.05172	0.05567	0.05962	0.06356	0.06749	0.07142	0.07535
0.2	0.07926	0.08317	0.08706	0.09095	0.09483	0.09871	0.10257	0.10642	0.11026	0.11409
0.3	0.11791	0.12172	0.12552	0.12930	0.13307	0.13683	0.14058	0.14431	0.14803	0.15173
0.4	0.15542	0.15910	0.16276	0.16640	0.17003	0.17364	0.17724	0.18082	0.18439	0.18793
0.5	0.19146	0.19497	0.19847	0.20194	0.20540	0.20884	0.21226	0.21566	0.21904	0.22240
0.6	0.22575	0.22907	0.23237	0.23565	0.23891	0.24215	0.24537	0.24857	0.25175	0.25490
0.7	0.25804	0.26115	0.26424	0.26730	0.27035	0.27337	0.27637	0.27935	0.28230	0.28524
0.8	0.28814	0.29103	0.29389	0.29673	0.29955	0.30234	0.30511	0.30785	0.31057	0.31327
0.9	0.31594	0.31859	0.32121	0.32381	0.32639	0.32894	0.33147	0.33398	0.33646	0.33891
1.0	0.34134	0.34375	0.34614	0.34849	0.35083	0.35314	0.35543	0.35769	0.35993	0.36214
1.1	0.36433	0.36650	0.36864	0.37076	0.37286	0.37493	0.37698	0.37900	0.38100	0.38298
1.2	0.38493	0.38686	0.38877	0.39065	0.39251	0.39435	0.39617	0.39796	0.39973	0.40147
1.3	0.40320	0.40490	0.40658	0.40824	0.40988	0.41149	0.41308	0.41466	0.41621	0.41774
1.4	0.41924	0.42073	0.42220	0.42364	0.42507	0.42647	0.42785	0.42922	0.43056	0.43189
1.5	0.43319	0.43448	0.43574	0.43699	0.43822	0.43943	0.44062	0.44179	0.44295	0.44408
1.6	0.44520	0.44630	0.44738	0.44845	0.44950	0.45053	0.45154	0.45254	0.45352	0.45449
1.7	0.45543	0.45637	0.45728	0.45818	0.45907	0.45994	0.46080	0.46164	0.46246	0.46327
1.8	0.46407	0.46485	0.46562	0.46638	0.46712	0.46784	0.46856	0.46926	0.46995	0.47062
1.9	0.47128	0.47193	0.47257	0.47320	0.47381	0.47441	0.47500	0.47558	0.47615	0.47670
2.0	0.47725	0.47778	0.47831	0.47882	0.47932	0.47982	0.48030	0.48077	0.48124	0.48169
2.1	0.48214	0.48257	0.48300	0.48341	0.48382	0.48422	0.48461	0.48500	0.48537	0.48574
2.2	0.48610	0.48645	0.48679	0.48713	0.48745	0.48778	0.48809	0.48840	0.48870	0.48899
2.3	0.48928	0.48956	0.48983	0.49010	0.49036	0.49061	0.49086	0.49111	0.49134	0.49158
2.4	0.49180	0.49202	0.49224	0.49245	0.49266	0.49286	0.49305	0.49324	0.49343	0.49361
2.5	0.49379	0.49396	0.49413	0.49430	0.49446	0.49461	0.49477	0.49492	0.49506	0.49520
2.6	0.49534	0.49547	0.49560	0.49573	0.49585	0.49598	0.49609	0.49621	0.49632	0.49643
2.7	0.49653	0.49664	0.49674	0.49683	0.49693	0.49702	0.49711	0.49720	0.49728	0.49736
2.8	0.49744	0.49752	0.49760	0.49767	0.49774	0.49781	0.49788	0.49795	0.49801	0.49807
2.9	0.49813	0.49819	0.49825	0.49831	0.49836	0.49841	0.49846	0.49851	0.49856	0.49861
3.0	0.49865	0.49869	0.49874	0.49878	0.49882	0.49886	0.49889	0.49893	0.49896	0.49900
3.1	0.49903	0.49906	0.49910	0.49913	0.49916	0.49918	0.49921	0.49924	0.49926	0.49929
3.2	0.49931	0.49934	0.49936	0.49938	0.49940	0.49942	0.49944	0.49946	0.49948	0.49950
3.3	0.49952	0.49953	0.49955	0.49957	0.49958	0.49960	0.49961	0.49962	0.49964	0.49965
3.4	0.49966	0.49968	0.49969	0.49970	0.49971	0.49972	0.49973	0.49974	0.49975	0.49976
3.5	0.49977	0.49978	0.49978	0.49979	0.49980	0.49981	0.49981	0.49982	0.49983	0.49983
3.6	0.49984	0.49985	0.49985	0.49986	0.49986	0.49987	0.49987	0.49988	0.49988	0.49989
3.7	0.49989	0.49990	0.49990	0.49990	0.49991	0.49991	0.49992	0.49992	0.49992	0.49992
3.8	0.49993	0.49993	0.49993	0.49994	0.49994	0.49994	0.49994	0.49995	0.49995	0.49995
3.9	0.49995	0.49995	0.49996	0.49996	0.49996	0.49996	0.49996	0.49996	0.49997	0.49997
4.0	0.49997	0.49997	0.49997	0.49997	0.49997	0.49997	0.49998	0.49998	0.49998	0.49998

Source: Engineering Statistics Handbook (National Bureau of Standards)

http://www.itl.nist.gov/div898/handbook/eda/section3/eda3671.htm

Answers to Chapter Questions

11-1.
a. Ho: P = .10
b. Ho: P > .10
c. 1.645
d. Reject Ho if Z > 1.645
e.

$$Z = \frac{\bar{P} - P}{\sqrt{\frac{P(1-P)}{n}}}$$

Substituting our problem values in the formula:

$$Z = \frac{.15 - .10}{\sqrt{\frac{.10(1-.10)}{100}}}$$

$$Z = \frac{.05}{\sqrt{\frac{.09}{100}}}$$

$$Z = \frac{.05}{\sqrt{.0009}}$$

$$Z = \frac{.05}{.03} = 1.67$$

f. Ho is rejected

11-2.
a. H_0: P = .4
b. H_1: P > .4
c. 2.33
d. Reject Ho if Z > 2.33
e.

$$Z = \frac{\bar{P} - P}{\sqrt{\frac{P(1-P)}{n}}}$$

Substituting our problem values in the formula:

$$Z = \frac{.475 - .4}{\sqrt{\frac{.4(1-.4)}{200}}}$$

$$Z = 2.17$$

f. Ho is not rejected

11-3

With a five percent level of significance the critical value of the test statistic would have been 1.645. The decision rule would have been "Reject Ho if Z > 1.645" The calculated Z of 2.17 would have indicated that the null hypothesis would be rejected.

11-4.
a. H_0: P = .7
b. H_1: P > .7
c. 2.33
d. Reject Ho if Z > 2.33
e.

$$Z = \frac{\bar{P} - P}{\sqrt{\frac{P(1-P)}{n}}}$$

Substituting our problem values in the formula:

$$Z = \frac{.775 - .7}{\sqrt{\frac{.7(1-.7)}{200}}} = 3.28$$

f. Ho can be rejected. If these results are stable, it appears that the candidate should win.

11-5.
a. H_0: Pd = Pn
b. H_1: Pd > Pn
c. 1.28
d. Reject H_0 if Z > 1.28
e. Calculations

The first step will be to find the pooled variance.

$$P_c = \frac{X_d + X_n}{n_d + n_n}$$

$$P_c = \frac{27 + 39}{100 + 200}$$

$$P_c = \frac{66}{300} = .22$$

$\overline{P}_d = 27 / 100 \qquad \overline{P}_n = 39 / 200$

$\overline{P}_d = .27 \qquad \overline{P}_n = .195$

Using the pooled variance and the formula provided earlier we can now calculate the value of Z, the test statistic.

$$Z = \frac{\overline{P}_d - \overline{P}_n}{\sqrt{\frac{P_c(1-P_c)}{n_1} + \frac{P_c(1-P_c)}{n_2}}}$$

$$Z = \frac{.27 - .195}{\sqrt{\frac{.22(1-.22)}{100} + \frac{.22(1-.22)}{200}}}$$

$$Z = \frac{.27 - .195}{\sqrt{\frac{.22(.78)}{100} + \frac{.22(.78)}{200}}}$$

$$Z = \frac{.075}{\sqrt{\frac{.1716}{100} + \frac{.1716}{200}}}$$

$$Z = \frac{-.075}{\sqrt{.001716 + .000858}}$$

$$Z = \frac{-.05}{\sqrt{.002574}}$$

$$Z = \frac{-.05}{.0507} = 1.48$$

g. Ho is rejected. 1.48 > 1.28

11-6.
a. $H_0: P_c = P_t$
b. $H_1: P_c \neq P_t$
c. 2.58
d. Reject Ho if Z > 2.58 or Z < - 2.58

e. Calculations

The first step will be to find the pooled variance.

$$P_c = \frac{X_d + X_n}{n_d + n_n}$$

$$P_c = \frac{27 + 39}{100 + 200}$$

$$P_c = \frac{66}{300} = .22$$

$\overline{P}_d = 27 / 100 \quad \overline{P}_n = 39 / 200$

$\overline{P}_d = .27 \quad \overline{P}_n = .195$

Using the pooled proportion and the formula provided earlier we can now calculate the value of Z, the test statistic.

$$Z = \frac{\overline{P}_c - \overline{P}_t}{\sqrt{\frac{Pc(1-Pc)}{n_1} + \frac{Pc(1-Pc)}{n_2}}}$$

$$Z = \frac{.33 - .275}{\sqrt{\frac{.3(1-.3)}{30} + \frac{.3(1-.3)}{40}}}$$

$$Z = \frac{}{\sqrt{\frac{.3(.7)}{30} + \frac{.3(.7)}{40}}}$$

$$Z = \frac{.055}{\sqrt{\frac{.21}{30} + \frac{.21}{40}}}$$

$$Z = \frac{}{\sqrt{.007 + .00525}}$$

$$Z = \frac{.055}{\sqrt{.01225}}$$

$$Z = \frac{.055}{.12068} = 0.50$$

g. Ho is not rejected.

11-7.

a. $H_0: P_a = P_b$
b. $H_1: P_a < P_b$
c. 1.645
d. Reject Ho if $Z < -1.645$
e. Calculations

The first step will be to find the pooled variance.

$$P_c = \frac{X_a + X_b}{n_a + n_b}$$

$$P_c = \frac{35 + 50}{500 + 500}$$

$$P_c = .085$$

$\bar{P}_a = 35/500 \qquad \bar{P}_b = 50/500$

$\bar{P}_a = .07 \qquad \bar{P}_b = .1$

Using the pooled variance and the formula provided earlier we can now calculate the value of Z, the test statistic.

$$Z = \frac{\bar{P}_a - \bar{P}_b}{\sqrt{\frac{P_c(1-P_c)}{n_1} + \frac{P_c(1-P_c)}{n_2}}}$$

$$Z = \frac{.07 - .1}{\sqrt{\frac{.085(1-.085)}{500} + \frac{.085(1-.085)}{500}}}$$

$Z = -1.70$

g. H_0 is rejected.

Chapter 12

Correlation

What this chapter will do for you.

As noted in the chapter describing Chi Square analysis, in an era before electronic calculators were common, there was a tendency to use techniques for which the burden of calculation was relatively light. This chapter briefly describes another analysis technique that can be used to explore possible cause and effect relationships, using calculations that are relatively simple. While Chi Square can be used for a hypothesis test, this technique, **correlation analysis** – while not providing a hypothesis test in the strict sense - can indicate whether or not two things have a negative or positive association. For example, larger amounts of water and fertilizer given to tomato plants – within the limits of what the plants need and can utilize - could be expected to be associated in a positive way with yields. More hours of television watching per week could be expected to be associated with student grades - in a negative fashion. Sometimes correlation will imply cause and effect, however you may not agree with an analyst that says A causes B. You might think it more logical that B might cause A. On the other hand A and B might both be due to something else; both being effects and neither a cause of the other. Or an apparent association might be totally accidental. (One research text notes that in Columbus, Ohio, the average monthly rainfall correlates nicely with the number of letters in the names of the months.*) In this chapter you will learn to calculate a coefficient of correlation ... and beyond that you can draw your own conclusions.

*(William K. Zikmund, Business Research Methods, 6th ed. Dryden)

The Coefficient of Correlation

This is a number that can range between plus one and minus one. Plus one would indicate perfect positive correlation. Negative one would indicate perfect negative correlation. Zero would indicate no correlation at all. To calculate the coefficient of correlation will use this formula. In the numerator, the average value of the X's is subtracted from each individual X and the average value of the Y's is subtracted from each Y. (The idea is that – when the summations are taken – if positive values cancel out negative values, there is little or no correlation.) In the denominator, the use of the term (n-1) and the standard deviations of the distributions of X and Y are used, and the calculation will come out between plus one and minus one and the sample size will not affect the size of r.

As you will see, a number of calculations are required, calculation of averages, subtractions, additions of the differences from those subtractions, multiplications, and calculation of standard deviations. The process can be somewhat laborious, but if we follow the procedure outlined below, build the table shown, and use values from the table in the formula, it is not extremely complicated.

Correlation Analysis begins with points on a graph of Cartesian coordinates. Let's use a simple example. Assume we have five experimental tomato plants which all receive the same amount of water but receive differing amounts of fertilizer. Ounces of fertilizer will be the independent variable (X), and plant height in inches the dependent variable (Y). The dependent variable is something that differs because of different inputs of the independent variable. The five plants are listed in the table below. Note: Data given in the problem appear in regular type. Results of calculations are italicized.

plant	X	Y	$X - \bar{X}$	$Y - \bar{Y}$	$(X-\bar{X})(Y-\bar{Y})$
1	1	4	*1-3 = -2*	*4-6 = -2*	*(-2)(-2) = +4*
2	2	5	*2-3 = -1*	*5-6 = -1*	*(-1)(-1) = +1*
3	3	6	*3-3 = 0*	*6-6 = 0*	*(0)(0) = 0*
4	4	8	*4-3 = +1*	*8-6 = +2*	*(+1)(+2) = +2*
5	5	7	*5-3 = +2*	*7-6 = +1*	*(+2)(+1) = +2*
					+9

First we need to calculate the average value of X. X will be 15 / 5 = 3.

Then we calculate the average value of Y. $\bar{Y}$ will be 30 / 5 = 6.

Then we will subtract the average of the X's from each X and put that value in the table in italics.

Then we will subtract the average of the Y's from each Y and put that value in the table in italics.

Then we will multiply across the rows, $(X - \bar{X})(Y - \bar{Y})$

Then we will add up the last column to get the sum of the $(X - \bar{X})(Y - \bar{Y})$ values to use in the formula below.

We will need also to calculate the standard deviations of the X's and the Y's. We can do that using the formula given back in Chapter One.

$$\text{Standard deviation} = \sqrt{\frac{\Sigma(x^2) - (\Sigma x)^2 / n}{n-1}}$$

To get values to plug into that formula we will build a table. First the X's.

X	X²		
1	1	$\Sigma x = 15$	$\Sigma x^2 = 55$
2	4		
3	9		
4	16		
5	25		
15	55		

$$\text{Standard deviation} = \sqrt{\frac{(55) - (15)^2/5}{5-1}}$$

$$\text{Standard deviation} = \sqrt{\frac{(55) - (225)/5}{4}}$$

$$\text{Standard deviation} = \sqrt{\frac{(55) - (45)}{4}}$$

$$\text{Standard deviation} = \sqrt{\frac{10}{4}}$$

$$\text{Standard deviation} = \sqrt{2.5}$$

Standard deviation = <u>1.58</u>

Now to find the standard deviation of the Y's.

We have one more number to calculate for the formula. The standard deviation of Y.

X	X²
1	1
2	4
3	9
4	16
<u>5</u>	<u>25</u>
15	55

$\sum x = 15 \qquad \sum x^2 = 55$

$$\text{Standard deviation} = \sqrt{\frac{(55) - (15)^2/5}{5-1}}$$

$$\text{Standard deviation} = \sqrt{\frac{(55) - (225)/5}{4}}$$

$$\text{Standard deviation} = \sqrt{\frac{(55) - (45)}{4}}$$

$$\text{Standard deviation} = \sqrt{\frac{10}{4}}$$

$$\text{Standard deviation} = \sqrt{2.5}$$

Standard deviation = <u>1.58</u>

Now we are ready to take values from the table and from these calculations and plug them into the formula for the coefficient of correlation, "r".

$$r = \frac{\sum(X - \overline{X})(Y - \overline{Y})}{(n-1) S_x S_y}$$

$$r = \frac{9}{(4)(1.58)(1.58)}$$

$$r = \frac{9}{9.9856}$$

$r = +.9012$ or $\pm .90$

As we might expect from looking at the X's and Y's in the table, the coefficient of correlation is almost plus one, indicating a very high degree of correlation.

Textbooks tell us that if we square the coefficient of correlation, we arrive at the **coefficient of determination**, which would tell us the percent of the variation in Y is caused by variation in X.

$r2 = .81$ We will assume that eighty one percent of the variation in height of the tomato plants is caused by variations in the amount of fertilized they receive.

Your analysis of correlation might be made stronger if you test the significance of the coefficient of correlation. This explanation may be a bit complex but let's give it a try. We have sampled a big population (possibly all the tomato plants that could possibly be raised and given this fertilizer). We found the correlation in our sample. Is it possible that in the big population the actual correlation is zero? That does not sound intuitively likely, but just with respect to numbers, if we took a huge number of samples we can be assured that – even though the correlation in a population is zero – we could find, in some small number of theoretically possible samples, samples with strong correlation. How likely is it that this might have happened in this case?

Surprise! We have arrived at yet another hypothesis test, this one a **t** test serving to answer the question posed above.

Ho: The correlation in the population is zero. $P = 0$
H1: The correlation in the population is different from zero. $P \neq 0$

$$t = \frac{r\sqrt{n-2}}{\sqrt{1-r^2}}$$

$$t = \frac{.90\sqrt{5-2}}{\sqrt{1-.81}}$$

$$t = \frac{.90\sqrt{3}}{\sqrt{.19}}$$

$$t = \frac{.90\ (1.7321)}{.4359}$$

$$t = \frac{1.5589}{.4359}$$

$$t = \underline{3.576}$$

We have calculated a **t**. Now how do we get a value from the **t** table to compare it with? We see because of the inequality sign that this is a two tail test. The use of five percent significance is fairly common, so we might as well use that. We originally had five data points. df = **n** – 2   So we will go into the t table at three degrees of freedom, at five percent for a two tail test, we find the critical value of the test statistic to be 3.182. (Recall: five percent significance for a two tail test will put .025 in each tail, so we read that column to get our critical value of **t**.)

Our calculated value of t exceeds the critical value of t so at the five percent level of significance we will reject the null hypothesis test that the correlation in the parent population is zero.

What have you learned and what comes next?

In this chapter you have learned how to compute the coefficient of correlation and the coefficient of determination and have learned how to do a t test of the coefficient of correlation. The next chapter, the last in this little tutorial book, will briefly address rank order correlation.

Values of Student's t distribution with degrees of freedom in column headings

Probability of exceeding the critical value

df	0.10	0.05	0.025	0.01	0.005	0.001
1.	3.078	6.314	12.706	31.821	63.657	318.313
2.	1.886	2.920	4.303	6.965	9.925	22.327
3.	1.638	2.353	3.182	4.541	5.841	10.215
4.	1.533	2.132	2.776	3.747	4.604	7.173
5.	1.476	2.015	2.571	3.365	4.032	5.893
6.	1.440	1.943	2.447	3.143	3.707	5.208
7.	1.415	1.895	2.365	2.998	3.499	4.782
8.	1.397	1.860	2.306	2.896	3.355	4.499
9.	1.383	1.833	2.262	2.821	3.250	4.296
10.	1.372	1.812	2.228	2.764	3.169	4.143
11.	1.363	1.796	2.201	2.718	3.106	4.024
12.	1.356	1.782	2.179	2.681	3.055	3.929
13.	1.350	1.771	2.160	2.650	3.012	3.852
14.	1.345	1.761	2.145	2.624	2.977	3.787
15.	1.341	1.753	2.131	2.602	2.947	3.733
16.	1.337	1.746	2.120	2.583	2.921	3.686
17.	1.333	1.740	2.110	2.567	2.898	3.646
18.	1.330	1.734	2.101	2.552	2.878	3.610
19.	1.328	1.729	2.093	2.539	2.861	3.579
20.	1.325	1.725	2.086	2.528	2.845	3.552
21.	1.323	1.721	2.080	2.518	2.831	3.527
22.	1.321	1.717	2.074	2.508	2.819	3.505
23.	1.319	1.714	2.069	2.500	2.807	3.485
24.	1.318	1.711	2.064	2.492	2.797	3.467
25.	1.316	1.708	2.060	2.485	2.787	3.450
26.	1.315	1.706	2.056	2.479	2.779	3.435
27.	1.314	1.703	2.052	2.473	2.771	3.421
28.	1.313	1.701	2.048	2.467	2.763	3.408
29.	1.311	1.699	2.045	2.462	2.756	3.396
30.	1.310	1.697	2.042	2.457	2.750	3.385
35.	1.306	1.690	2.030	2.438	2.724	3.340
40.	1.303	1.684	2.021	2.423	2.704	3.307
45.	1.301	1.679	2.014	2.412	2.690	3.281
50.	1.299	1.676	2.009	2.403	2.678	3.261

Source: Engineering Statistics Handbook (National Bureau of Standards)

http://www.itl.nist.gov/div898/handbook/eda/section3/eda3672.htm

Chapter 13

Rank Order Correlation

What this chapter will do for you.

It is not common but it is possible that a research project will yield data in ordinal form. Here, the ability to conduct an analysis using rank order correlation can be useful. While not a hypothesis test, such an analysis can shed light on apparent relationships, or lack of them, between two entities or phenomena for which the researcher can gather ordinal data.

Rank Order Correlation

Chapter 11 discussed the sample coefficient of correlation, which could be used for interval and ratio data. **Rank Order Correlation** analysis can compare things such as rankings of job applicants by interviewers and later rankings of those same employees once on the job or rankings given job candidates by two different interviewers. The measure is called Spearman's coefficient of Rank Correlation after its originator, Charles Spearman, a British statistician. The formula for its calculation is provided below.

$$r_s = 1 - \frac{6 \sum d^2}{n(n^2 - 1)}$$

In this formula the variable "d" is the difference between ranks for each pair and "n" is the number of paired observations.

Let's go through a simple example to see how this analysis is performed. In this example Smith and Jones have each evaluated five candidates for jobs in the engineering department of Precision Products. Using grade point average coupled with a subjective evaluation of work ethic, perseverance, initiative, and ability to work as part of a team, both Smith and Jones have ranked these candidates. We wish to evaluate the extent to which these ranking agree. (Note: This is a sample of a larger population of rankings which Smith and Jones might do over an extended period of time. The purpose is to provide insight as to the likely consistency of ratings between these two evaluators in that larger population of rankings.)

Candidate	Smith Ranking	Jones Ranking	Difference	Difference Squared
A	2	3	-1	1
B	5	5	0	0
C	3	4	-1	1
D	4	1	3	9
E	1	2	-1	1
			$\sum d = 0$	$\sum d^2 = 12$

$$r_s = 1 - \frac{6 \sum d^2}{n(n^2 - 1)}$$

$$r_s = 1 - \frac{(6)(12)}{5(5^2 - 1)}$$

$$r_s = 1 - \frac{72}{5(25 - 1)}$$

$$r_s = 1 - \frac{72}{120}$$

$$r_s = 1 - .60$$

$$r_s = .40$$

Since this is an estimate of the correlation in the overall population of rankings by these to evaluators, should we assume that it does really show relatively strong correlation in the population.

Here, a hypothesis test is required.
Ho: The rank correlation in the population is zero.
H1: There is a positive association among the rankings given by Smith and Jones.

$$t = r_s \sqrt{\frac{n-2}{1-r_s}}$$

$$t = .40\sqrt{\frac{5-2}{1-.40}}$$

$$t = .40\sqrt{\frac{3}{.60}}$$

$$t = .40\sqrt{5}$$

$$t = .40\sqrt{2.2361}$$

$$t = .8944$$

The next step is to get a critical value of with which to compare our calculated value of t. We had five pairs of rankings so we will use 5-2 or three degrees of freedom. Five percent significance is a commonly used number so we go into the table at five percent in one tail column (since this is a one tail test) and find a value of 2.353. Our calculated value does not exceed that critical value of the test statistic so we would conclude that the null hypothesis cannot be rejected and the alternate accepted – there is a positive association between the rankings given by Smith and Jones.

What have you learned and what comes next?

From this chapter you have learned how to do rank order correlation, which could be useful if your research involves ordinal data.

From this little tutorial book, the writer hopes that you have (1) overcome undue apprehension that may have existed about the difficulty of applying simple hypothesis tests, (2) gained mastery over each hypothesis test presented here, (3) acquired the ability to determine whether or not the data from your research project can lend itself to a hypothesis test, and, if so, which test would be most appropriate, and (4) put yourself in a position to conduct whatever test you may choose, accurately and with confidence.

*****

A problem to practice on.

You are a coach for a professional football team. You desperately need a so-called "go-to guy" to play wide receiver. Nothing fell your way in the draft but you have a list of ten undrafted graduating college players who might fill the bill. None are very big but all are reasonably fast and had good numbers in college. With the thought that you might find a player too small to play college ball and, once he did, too small to be a pro, who nevertheless would turn out to be outstanding - you ask your receivers coach and a university psychologist to interview each of the ten and rank them with respect to work ethic, courage, determination, intelligence, and character. You intend to hire one or more as free agents. Before you take that step you wish to compare the rankings give by the receivers coach and the psychologist.

Player	Receivers Coach	Psychologist	Difference (d)	d²
1	9	8	—	—
2	5	2	—	—
3	3	1	—	—
4	6	6	—	—
5	1	3	—	—
6	10	10	—	—
7	4	5	—	—
8	7	9	—	—
9	2	4	—	—
10	8	7	—	—

$$r_s = \frac{6 \sum d^2}{n(n^2 - 1)}$$

To test the significance of rs

Ho:

H1:

$$t = r_s \sqrt{\frac{n - 2}{1 - r_s}}$$

Player	Receivers Coach	Psychologist	Difference (d)	d²
1	9	8	*1*	1
2	5	2	3	9
3	3	1	2	4
4	6	6	*0*	*0*
5	1	3	*-2*	4
6	10	10	0	0
7	4	5	*-1*	1
8	7	9	*-2*	4
9	2	4	*-2*	4
10	8	7	*1*	*1*

$\sum d = 0 \qquad \sum d^2 = 28$

$$r_s = 1 - \frac{6 \sum d^2}{n(n^2 - 1)}$$

$$r_s = 1 - \frac{(6)(28)}{10(10^2 - 1)}$$

$$r_s = 1 - \frac{168}{10(99)}$$

$$r_s = 1 - \frac{168}{990}$$

$$r_s = 1 - .1697$$

$$r_s = .8301$$

To test the significance of rs
Ho: The rank correlation in the population is zero.
H1: There is a positive association between the rankings.

$$t = r_s \sqrt{\frac{n-2}{1-r_s}}$$

$$t = .8303\sqrt{\frac{10-2}{1-.8303}}$$

$$t = .8303\sqrt{\frac{8}{.1697}}$$

$$t = .8303\sqrt{47.1420}$$

$t = .8303\ (6.8660)$

$t = \underline{5.708}$

At 10 – 2 or 8 degrees of freedom, at five percent, the ttable gives us a critical value of 1.860. Since our calculated t is greater than that we will reject the null hypothesis.

Made in the USA
Middletown, DE
09 March 2020